TECHNOLOGICAL LANDSCAPES
RICHARD ROGERS

CRD DOCUMENTS
A SERIES OF POLEMICS, MANIFESTOS, ESSAYS AND LECTURES
ON COMPUTER RELATED DESIGN

GENERAL EDITOR GILLIAN CRAMPTON SMITH
SERIES EDITOR GILES LANE

FIRST PUBLISHED IN 1999 BY THE ROYAL COLLEGE OF ART
COMPUTER RELATED DESIGN RESEARCH STUDIO
KENSINGTON GORE
LONDON SW7 2EU
www.crd.rca.ac.uk/research

© 1999 THE ROYAL COLLEGE OF ART & RICHARD ROGERS

ALL RIGHTS RESERVED. NO PART OF THIS BOOK MAY BE REPRODUCED IN ANY
FORM BY ANY ELECTRONIC OR MECHANICAL MEANS, INCLUDING PHOTOCOPYING,
RECORDING, OR INFORMATION STORAGE AND RETRIEVAL WITHOUT PERMISSION
IN WRITING FROM THE PUBLISHER

BOOK DESIGNER PAUL FARRINGTON

PRINTED IN THE UK BY GEOFF NEAL LITHO, MIDDLESEX
TYPE MATRIX, TRIPLEX, INTERSTATE & POPLAR
PAPER TEXT – JAMES McNAUGTON®, 140GSM CYCLUS OFFSET
COVER JAMES McNAUGTON®, 280 MICS CHALLENGER PULPBOARD

BRITISH LIBRARY CATALOGUING-IN-PUBLICATION DATA:
A CATALOGUE RECORD FOR THIS BOOK IS AVAILABLE AT THE BRITISH LIBRARY

ISBN: 1 874175 28 4

THIS PUBLICATION HAS BEEN GENEROUSLY SUPPORTED BY
INTERVAL RESEARCH CORPORATION

Royal College of Art
Postgraduate Art & Design

TECHNOLOGICAL LANDSCAPES
RICHARD ROGERS

RCA **CRD** RESEARCH

CONTENTS

PRELUDE	**08**
THE LAWN AS TECHNOLOGICAL LANDSCAPE	**12**
RECALLING FORDISM	**20**
THE AGE OF SPEED	**26**
MAKING USABLE HISTORY FROM SYSTEMS BATTLES	**34**
SUBLIME CONTROL: LEGACIES OF THE ATOMIC & ANTI-ATOMIC AGES	**44**

TECHNOLOGICAL NATIONALISM & RELEVANT PAST FUTURES FOR THE COLONISATION OF MARS	54
FLIRTING WITH THE GHOST IN THE MACHINE	62
SEVENTIES NOW! MODERN & POSTMODERN TECHNOLOGICAL DESIGN	70
LISTEN, CITIZEN	76
TRUTH IN THE MUSEUM & THE REFLEXIVE EXHIBITION	84
DESIGN FOR ALL: TECHNOLOGIES FOR THE SITUATIONALLY CHALLENGED	92

PRELUDE

PERHAPS the first technological landscape was the Jeffersonian..The 'middle landscape' would be a pleasant combination of the natural and the industrial. Against the backdrop of wild nature, the train would whistle without shrieking and factory smoke would wisp without billowing.**(01)** Tidy 'light' industry, as we call it now, would scape the virgin land. The middle landscape would be the winning marriage of technology and nature, economically, culturally and aesthetically.

The planned American mill towns of the 1830s and 1840s, such as Lowell and Lawrence, Massachusetts, briefly realised that mini-utopian ideal, only to see themselves gradually become what they were expressly designed not to be – English-style 'coketowns'. A century later, when visitors queued for the General Motor's Futurama pavilion at the 1939 World's Fair in New York city, the pervading image of America's small and medium-sized cities was of a Massachusetts mill town, or the literal Pennsylvanian coketown, gone horribly wrong and still in severe depression. ('Smog', an amalgamation of smoke and fog, was coined in Pittsburgh, Pennsylvania). General Motor's city of the future – green, airy, slum-free and car-friendly – would replace the smog-ridden dystopia; it was Eisenhower's interstate highway system together with brutalist urban renewal that materially realised this image of the new technological landscape.

Particularly in the early 1970s the ecological movement (among others) challenged both that concept and reality of the futuristic landscape built for mobile, high-rise living. In their attempts to meet the challenge, planners envisioned two kinds of hybrids to tidy the dreary modernist picture. The beloved pastoral-technological landscape idea, inherited from the middle landscape, assumed high- and low-tech versions; one powered by nature (solar or water), the other powered by another allegedly 'clean' source, nuclear energy. Such hybrids, devised to improve the quality of life through designing productive and aesthetically pleasing techno-naturalesque spaces, lie behind the notion of the technological landscape.

Of course, technological landscapes of a kind have been with us all along in the guise of the garden, the park and the zoo. These more conventional 'manufactured' landscapes (which also include the wildlife corridor) may require technology for their demarcation and upkeep (for their 'landscaping'), but they are built for and run on nature. An archetypal technological landscape these days would be the wind field, the machine garden, the water park, even the skyline – technological spaces couched in natural, or naturalesque terms. These new 'design-scapes' are built for technology, and run on technology. Nowadays, to save nature, spur markets, recreate communities or encourage contact between distinct ethnic and cultural groups, designers are recasting neighbourhoods, cities, regions, countries and even transnational spaces as design-scapes, often borrowing from the history of the imagined future.

Since technological landscapes are devised and designed – not given by nature but inherited from history – choices as to the most desirable image of the future may be made. Indeed, the history of technological culture (the hybrid historiographical perspective adopted in this book) demonstrates the possibility of pursuing alternate paths, of choosing between relevant past futures. A map of past and present roads taken and not taken, the history of technological culture invites rereadings to inform the selection of the technological landscapes of our day.

The following essays are adapted from lectures given at the University of Amsterdam, the Royal College of Art (London), the Netherlands Design Institute and newMetropolis Science and Technology Center, Amsterdam. They derive from 'technological culture', a body of thought informed by science and technology studies, the sociology of technology, the history of technology and culture, and the study of design. Emerging in Amsterdam academic and public intellectual culture over the past decade, this body of thought departed from the premise that "technology is the dominant style of thinking in western culture... [It] has become a 'natural' part of our lives."**(02)**

A capsule history of western technological culture, far more associative than exhaustive, would set out the principles, visions, ideals and ideologies which have underpinned that style of thinking. It would examine the forms of social organisation that the technologies bring with them and that we now experience as second nature. It would explore how the technologies and their associated cultures were accepted in the first place, how they were promoted and assessed as they were under development, and first introduced to the public at large. It would point out the wider cultural reactions and social movements they spawned. It would inquire into how one technology or technocultural belief system won out over another viable alternative. In all, it would provide a kind of history of technocultural ideas, inviting the reader to draw analogies between past and present choices.

The spinning mill, the train, the telephone, the atomic bomb, the rocket, the computer — the initial starting point for many of the essays is the western culture that fostered, and unfolded around, these major technologies. Beginning with the contemporary context, each essay strives to lay out the 'transhistorical' nature of technocultural ideas, the recurrence of thought and design style over time, and in our times. The Fordist will, the infatuation with speed, the struggle between competing systems, the fascination for the technological sublime, the imagination of the future, the spirit inside the machine, the modernist aesthetic, the civic ideal, the communication of risk to the public, the redesign of the body — each of these themes emerge (more or less chronologically) in the respective essays. The book may be read as a short history of technological culture from the railways to the Internet.

Taken together the essays make a series of recurring points for technologists and critics. Much revolves around anticipative historical comparison, which NASA understood in the mid-1960s, when the agency invited historically-minded social scientists to study whether the railroad would be an appropriate analogy for promoting space travel.**(03)** More recently the U.S. Government chose the highway system as the appropriate analogy for promoting the Internet.

Historical comparison, with imagery of previous technological landscapes, fires the imagination. It is also the stuff of argument and defence for an idea or a project. In these essays attention is paid to how technologies have been argued for and against in the past. In making a well-reasoned argument for a new product or system, principally three types of historical analogy are employed: comparison between individual technologies past and present, between explanations for technological and social change past and present, and between protests against individual technologies past and present. Knowledge of previous reactions furnishes a guide for interpretations of the present, it also aids in anticipating and answering criticism. Whether in (nuanced) praise or fear of a new technology, the choice of the apt and the avoidance of the spurious analogy with past technologies would be a useful lesson.

Finally, the new landscape, whether thought of in terms of manufactured nature or the technological landscape, connotes a sense of viewer appreciation. In a technological landscape, it is the organisation of the technology that appeals; in the manufactured landscape it is the organisation of nature. An appreciation for the aesthetic of the technological landscape and 'new nature' (as the manufactured nature aesthetic may be called) permeates much of this book.**(04)**

(01)
Marx, L., *THE MACHINE IN THE GARDEN*, Oxford University Press, New York (1964).

(02)
Schwarz, M. & R. Jansma (eds.), *DE TECHNOLOGISCHE CULTUUR*, Uitgeverij de Balie, Amsterdam (1988) p. 1.

(03)
Mazlish, B. (ed.), *THE RAILROAD AND THE SPACE PROGRAM: AN EXPLORATION IN HISTORICAL ANALOGY*, MIT Press, Cambridge (1965).

(04)
The Dutch Pavilion for EXPO 2000 in Hannover, Germany is conceived as a survey of new nature.

01
THE LAWN AS TECHNOLOGICAL LANDSCAPE

The Lawn as Technological Landscape

READERS of *HARPER'S* magazine, the American monthly, are familiar with its provocative and sometimes blackly humorous 'Index', in which 'factoids' are artfully juxtaposed to make a series of damning statistical statements, often about the American culture of consumption. To illustrate the pride of place assumed by the green, velvety outdoor carpet in American society, the following index has been culled from the pages of *THE LAWN* by Virginia Jenkins (1994).**(01)**

Estimated number of lawns in the U.S. in 1989, in millions: **45**. *Estimated amount of space covered by lawns in the U.S. in 1989, in millions of acres:* **30**.

Estimated annual revenue of U.S. lawncare industry in 1987, in billions of dollars: **2.8**. *Estimated amount of money spent tending American lawns in 1989, in billions of dollars:* **5.3**.

Recommended amount of nitrogen fertiliser applied to an American lawn in 1940, in pounds: **1**. *Recommended amount in 1970:* **8**.

Approximate number of golf courses in the U.S in 1902: **1,000**. *Number of golf courses built in the U.S. in 1964 and 1965:* **1,000**.

Estimated amount of chemicals applied to American lawns in 1989 in millions of pounds: **70**. *Maximum estimated annual increase in amount of chemicals applied to American lawns since 1989, in percent:* **8**.

Estimated amount of water required in summertime to maintain a 25-by-40 foot lawn, in thousands of gallons: **10**. *Estimated annual water consumption on lawns in some western U.S. states in 1990, as a percentage of total water consumption:* **67**.

*Maximum estimated percentage of American landfills consumed by yard waste***(02)**, *in 1989:* **50**. *Number of states having or considering laws restricting the dumping of yard waste, in 1991:* **34**.

Number of electric-powered mowers produced in the U.S. in 1989, in millions: **5**. *Number of American companies producing manually-operated (push) mowers, in 1987:* **1**.

Number of people treated in American hospital emergency rooms for lawn-mower related injuries, in 1989: **60,000**. *Average annual number of mower-related deaths in the U.S., in the 1980s:* **100**.

Over the past century, the phenomenal growth in the number and expanse of lawns in the United States is attributed to the technoscientific capacity to grow and maintain green swards as well as to the potency and endurance of the aesthetic symbolism of the perfect lawn. A well-kept front lawn is the outdoor symbol of indoor stability, an indication of domestic tranquillity. All over the country, even in the mòst arid states, residents of homes with untidy lawns or no lawn at all are considered lazy, morally derelict, even unAmerican. As many foreign residents of the U.S. have learned, not only is neighbourly peer pressure high to tend the grass, but many local communities have statutes punishing offenders for unkempt yards. First-time American homeowners are especially aware of appearances, so the first step in making a home is making a lawn and that's Dad's job. Putting in and regularly tending a lawn keeps Dad fit, raises the value of the property and signals civic pride, especially on Flag Day and other patriotic holidays, when the Stars and Stripes is appropriately framed by a picture-perfect home and lawn.

The front lawn is a showpiece to be admired by the neighbours, before they're invited over for a glass of tonic and sandwich wedges without the crusts. Once inside, the neighbours are ushered into the indoor equivalent of the front lawn, the immaculate, rarely trespassed living room, not to be confused with the family room or the den, the indoor extension of the backyard, where children play by day and parents watch television in the evening. The backyard, like the family room, is there to be used – for 4th of July barbecues, wiffle ball games and Dad's attempt at a hardy vegetable garden. The toolshed, the swimming pool, the birdbath and the weeping willow tree are also in the backyard or, perhaps, a recessed side yard. Only the neatest and firmest accoutrements belong in the front yard, like cigar store indian shrubbery, the flagstone walkway and the mailbox, ideally with the red signal flag raised. This is the archetypal American home, Political Correctness disclaimers and Tom Waits not included.

Fittingly, George Washington had America's first publicised lawn. Modelled after an English country estate's rolling (and bowling) green, the lawn in front of Mount Vernon, as portrayed in a famous eighteenth century lithograph, was characteristically traversed by grazing sheep, dogs and other natural ornaments. By the 1970s almost every American suburban homeowner, and many urban dwellers besides, were coveting and attempting to cultivate the perfect lawn, a green, velvety carpet – perfectly trimmed, edged and weed-free. Sheep had been replaced by power mowers and dogs by artificial pink flamingos. Perhaps the crowning achievement of the perfect lawn has been its Seventies-style simulation: a even bed of set concrete painted lawn green. Since the rise of the synthetics industry, Americans have dignified their front yards with green-outdoor carpeting, Neo-turf (green vinyl), Permagrass (green plastic fibres), Polyloom II ('grasslike surfacing'), TailorMade lawn (cellulose), Astroturf and other 'hassle-free' real grass substitutes. Just as naturally unnatural was the great horticultural event outside of Detroit at soccer's World Cup '94, when special 'natural'

THE LAWN AS TECHNOLOGICAL LANDSCAPE

turf was laid inside the enclosed Silverdome for the opening game. While the 1980s and early 1990s have witnessed the emergence of the chemical-free Natural Lawn Movement, an apt oxymoron (for there is nothing natural about a lawn), the "vital aesthetic component" of the American landscape seems to be as difficult to dislodge as the automobile, Middletown's measure of man (p.180).

How, then, have proverbial white, middle-class suburban Americans become so 'obsessed' with increasingly unnatural home lawn aesthetics between the days of George Washington and the natural lawn movement? The answer lies in the cultural confluence of lawncare equipment industrial advertising, home and garden magazines, golf, garden clubs, government-leisure industry research alliances, golfing presidents and 'consumption communities'. All have had a hand in manufacturing the "ability and desire to grow and tend lawn grasses", which are not indigenous to North America (p.183). The constitution of a lawn, both technoscientifically and aesthetically, has been progressively perfected over the years. Akin to rising standards of household cleanliness and family and personal health and hygiene,[03] notions of the perfect lawn have undergone transformations from rolling greens before of the turn-of-the-century to velvety fairway turfs and from seasonal lawns to year-round green swards and artificial turfs and grasses. Except during the World Wars when, much to the chagrin of lawncare specialists, many Americans mangled the aesthetic by growing vegetables in the front lawn 'victory garden', perfectionism has increased, as have the amount of money spent, grass seeds imported and scientifically manipulated, equipment manufactured and procured, chemicals invented and applied, refuse produced and discarded and water, gas and electricity consumed. One of the main players in the historical growth of the lawn aesthetic, the United States Golf Association (USGA), founded in 1894, now has the pièce de resistance surrounding its headquarters in New Jersey: "acres of weed-free, insect-free, disease-free grass of a uniform colour and height, made possible by eight decades of research." (p.151)

Manufacturing consumer demand through power politics has been a subject of the social study of science and technology for some time. Success in the marketplace is sometimes a product of prior political alliance-building between developers, producers and distributors of the components of a technological system, who themselves are supposedly reacting to consumer demand — a virtuously circular argument to vindicate vested interest.[04] Especially in gender & technology, we also read about the concomitant project of imbuing technology with powerful symbolism through advertising, industry-sponsored community programs and other industrial outreach.

Gardening and lawncare advice was available for the well-to-do in the latter half of the nineteenth-century, as naturalesque urban park designers like Olmsted and the more formalist stylists of the fledging City Beautiful Movement lent credence to the beauty and necessity of lawncare for residents of the old estates, already accustomed to the English aesthetic, and the new late nineteenth-century suburban communities, dubbed 'Parks', like

THE ELECTRIC LAWNMOWER BY SOUTHERN CALIFORNIA EDISON. "IN FACT, THE ONLY TYPE OF LAWNMOWER THAT'S MORE ENVIRONMENTALLY-FRIENDLY IS A MANUAL LAWNMOWER."

West Orange New Jersey's Llewellyn Park of Thomas Edison fame.**(05)** The aesthetic caught on, not necessarily owing to a civilising process of imitation by the would-be bourgeoisie (as the sociologist Norbert Elias has described), but also, in part, by design.**(06)** City beautification projects of the first decade of the twentieth century encouraged civic participation. The general public was urged to pitch in for the good cause, and public lawn projects and contests were organised by progressive clubs, affluent women's groups and large companies, which promoted 'welfare work' as an alternative to labour unrest.

Beginning in the 1910s, the United States Department of Agriculture, which had been applying the fruits of its research to federal lawns, and the USGA forged alliances for scientific research on new grasses and hybrids. Funded in whole or in part by the Golf Association, Government Agriculture Experiment stations, beginning in the 1920s, produced new hearty seeds and turfs for immediate application on the links of America, many of which were now public or municipal and thus open to motoring Americans with weekends free for leisure activity. A cash crop, grass was big business as was its biggest customer, the golf industry, which according to a 1926 magazine article, was valued at a billion dollars.

Putting greens and fairways require management: fertiliser, grass seed, lawn equipment, mowers and tenders. In their advertising, Toro and Scott, then profiting from the interest in golf and still large lawncare equipment manufacturers, compared the golf course to the suburban lawn. So did the great catalogue merchandisers, Sears Roebuck and Montgomery Ward, and golfers returning from their eighteen holes, together with their catalogue-consumer families, began rethinking their landscape design. Gradually, "the image of a velvety carpet became ubiquitous in twentieth-century advertising" (p.80), as lawncare company publications, mail order catalogues and home and garden, women's and general interest magazines promised the lawn-tending homeowners that they'd be the envy of every neighbour. While lawncare advertising took many forms, generally speaking wives were invited to purchase lawn beauty supplies to improve appearances, such as fertiliser, while their husbands were tempted by the newest power technologies for gas and electric mowers. Women were encouraged to prod their husbands to do the real work, then take pride in the beautiful landscape aesthetic. Men would be reminded of the long fairway of a par 5 hole. Even in power mower advertising directed at both men and women, the assumption remained that Dad would mow the lawn. If he couldn't get around to it one weekend, the power mowers were supposedly easy enough for Mom to use. Jenkins suggests that this sort of phraseology discouraged women from even trying, thereby reinforcing their role as decorator of the interior as well as the exterior of the house. Mom pictures the perfect arrangements and Dad moves the furniture and pushes the lawnmower. Only much later would the lawncare service industry partially liberate Dad from yard work.

American lawns and the lawncare industry also suffered from the war, although the Department of Agriculture urged Americans to restrain from "spading up front lawns... to grow vegetables. Too much of this... was done in the last war, and most of it paid pretty poor dividends" (p.95). The lawncare industry, according to one advertiser, was "standing by" and waiting for 'V Day'. As the lawn aesthetic returned gracefully after the war, frustration set in, owing to Dad's inability to root out weeds and pests and achieve the standard velvety, green carpet, now of a single type of grass. Enter the chemical industry, which had been experimenting with inorganic fertilisers in the 1930s, and entire lawncare regiments advertised by industry, like Scott's "Whole Lawn Program", which included power mowers, equipment, seed, food, and chemicals to feed and care for the lawn. By the mid-1950s, the power mower had become a fixture on the American lawn, and 'war' was raging between the male homeowner and any number of new pests (Japanese beetles, ants, animals, insects, worms, diseases, weeds) now residing in the novel grasses. If Dad lost a battle, there were chemicals at his disposal to kill every living thing in the front yard and start afresh. Tested by Government Agricultural Stations during the war and injected in weed guns, weed bombs and similarly belligerent lawncare supplies from the end of the war until it was banned 1972, the 'killer of killers', DDT, as well as other herbicides and pesticides, made its ignominious presence felt across the lawns and homes of America. This is a capsule pre- and proto-history to Rachel Carson's *SILENT SPRING*, which for all its impetus to the environmental movement has not impeded Chemlawn and similar lawncare corporations from posting sign after sign indicating "Another Satisfied Customer" as well as "Please Stay Off the Grass Until Dry".

Attempts were made to persuade former President George Bush, another in a long line of golfing presidents beginning with Woodrow Wilson, to rip out the White House lawn and landscape the tone-setting terrain in the style of a meadow, wetland, vegetable garden or fruit orchard. Rejoinders from lawncare advocates revolve around the environmental and psychological benefits of the front lawn; it reduces noise and air pollution, absorbs heat and glare, processes carbon dioxide, beautifies the surroundings and affords peace and serenity. Jenkins concludes with the natural lawn movement, members of which are said to be considered 'organic cultists', and an invitation to social scientists: "A new landscape aesthetic is a cultural creation, and it remains to be seen whether the environmental movement in this country can enlist as potent a group of supporters and teachers for the twenty-first century as the lawn industry, the Garden Club of America, the U.S. Golf Association, and the U.S. Department of Agriculture did during the twentieth century." (p.187) Initiatives are afoot to study and challenge the automobilismus and birthright mobility of Americans, but what of strategies to 'imperfect' the lawn?

This essay was published previously as "More Work for Father and Son: The Problems of the Perfect Lawn in the U.S.A", in *EASST REVIEW*, 14, 2, 1995, pp. 3-6.

(01)
This piece is based on Jenkins, Virginia Scott, *THE LAWN, A HISTORY OF AN AMERICAN OBSESSION*, Smithsonian, Washington, DC 1994. The page numbers of the statistics are 187, 187; 168, 181; 142, 142; 31, 60; 186, 186; 186, 172; 173, 173; 112,179; and 114,115, respectively. The other page numbers in parentheses throughout the piece also refer to Jenkins.

(02)
See the biodegradibity studies by a team of garbologists in Rathje, W. & C. Murphy, *RUBBISH! THE ARCHAEOLOGY OF GARBAGE*, HarperPerennial, New York, 1992.

(03)
Cowan, R.S., *MORE WORK FOR MOTHER*, Basic Books, New York, 1983.

(04)
See Braun, H.-J., "Introduction", Symposium on 'Failed Innovations', *SOCIAL STUDIES OF SCIENCE*, 22, 1992, pp. 213-230. See also Wajcman, J., *FEMINISM CONFRONTS TECHNOLOGY*, Polity, London, 1991.

(05)
See Wilson, W., *THE CITY BEAUTIFUL MOVEMENT*, Johns Hopkins Univ. Press, Baltimore, 1989.

(06)
Elias, N., *THE CIVILISING PROCESS, VOL. I, THE HISTORY OF MANNERS*, Oxford, Blackwell, 1978.

02
RECALLING FORDISM

IN early 1998 a meeting was organised at a pub in London on the topic 'digital artisans'. The digital set — hipsters and suits alike — arrived in goodly numbers. The meeting stopped short of being a rally; it was an interesting attempt to organise an identity, a movement even, of the digital set. A historical comparison was drawn — between the digital designers and makers of today and the European and British artisans of the eighteenth and early nineteenth centuries as skilled craftsmen. Further analogies were trotted out. Just as today's digerati do not don neckties, the artisans did not don britches (they were sans culottes). Today's designers, in deference to workers, often wear blue jeans just as the Dadaists in the 1920s wore overalls in deference to the skilled mechanics of the Machine Age.

What was not said was that both the artisans of yesteryear and today's digital designers work in non-Fordist regimes which is why the call to organise, to associate, to collectivise, to price-fix and to wage-fix, fell upon deaf ears. There was little hope for today's digital artisans to answer the call to action for at least two reasons. First, historically, economic upswings have not been the wellsprings of critical movements calling for a change to the dominant mode of manufacture. On the contrary, periods of economic prosperity have gone hand in hand with technological hubris. Early Victorian prosperity saw the growth of enthusiastic science fiction; the fin-de-siècle saw the development of Futurism and infatuations with technologies of speed, the 1920s and 1950s saw technological consumption advertising, and the 1990s Internet style magazines and wearable technology fashion shows.

But, more importantly, there is currently no recognised alternative to post-Fordist flexibility and, barring the Unabomber, no recognised protest against that mode of work.**(01)** Nowadays, the general view is that creativity and individual initiative in the workplace are not being squashed by some variation of modernism, as it was when the Luddites rose up against mechanisation in the early nineteenth century and when unions organised against mass production and rationalisation in the early 1900s, or when a colourful spectrum of groups protested scientification, cold-war bureaucratisation and automation in the 1960s and early 1970s. Even the 1980s hackers have vanished from the cultural landscape, having shifted from illegal cracking to corporate cracking protection. Project workers, as the avant-garde digital artisans may be more aptly called, do not appear to be subjected to the traditionally disciplined, paternalistic workaday regimes of yesteryear. They do not organise, for their time, at least between project deadlines, appears to be their own. 'Flexitime' also allows them to retain some sense of personal autonomy, of having one's own time.

This was not the case in the early nineteenth century of the Luddites, or earlier this century with the unions. Luddism, as Kirkpatrick Sale (a Unabomber commentator) explains in *REBELS AGAINST THE FUTURE* (1996), was a local machine-breaking movement which arose in the heart of industrial Britain in the early nineteenth century. The machine breakers were reacting not only to mechanisation but to the replacement of a way life based on nature by another based on technology, the bell. Skilled craftsmen — initially handloom weavers, and combers and dressers of wool, working in cottages and small workshops — were put out of their traditional work as well as their rhythm of life by the ascendancy of the spinning mill (the factory) with its omnipotent and omnipresent bell. It was an authoritarian, paternalist regime. In American mill towns from the 1830s until the first successful labour strikes, the day's first bell would ring as early as four-thirty in the morning. A mill girl of Lowell Massachusetts wrote: "Up before day, at the clang of the bell — and out of the mill by the clang of the bell — into the mill, and at work, to the obedience of that of the bell — just as though we were so many living machines." In mill towns like Lowell, all the mill clocks would be rung in unison. Wake at four-thirty, work at four-fifty, breakfast, dinner, quitting time and evening curfew at ten o'clock — all would be regulated by the bell, every day of the week.

Robin Murray has called the factory workers' trade-off a 'Faustian bargain' — giving up a way of life on the farm or in cottages, giving up skills and mental faculties — for a higher wage.**(02)** The workers spent the money they earned on factory-owned housing and on factory-made products keeping the system firmly in place. In exchange for higher wages, they sold their nature, their time, to the regime.

The length of the working day, the higher and higher speeds of the machines, the stuffy, hot, noisy and dangerous factory working conditions and the wages — that Faustian bargain — these were the issues for the workers. Instead of breaking the machines, Luddite-style, that fed themselves and their families, they organised and took their grievances to the management, to the press and the state legislatures. The long fight — often bloody — ultimately led to the call for (what in Europe is called) the social democratic ideal of 8-8-8: eight hours work, eight play and eight sleep. Had the digital artisan ralliers in London mentioned the length of the workday and new bodily injuries of the information age, they might have had more reason to prepare a movement for when the economy takes a downturn.

Nineteenth-century labour gains should not be overestimated, however. In industrial New England factory owners not only would break the strikes violently, but would replace one wave of rebellious workers after another with new blood, each more pliable and more dependent on the wages. Thus the rural mill workers of the 1830s and 1840s were replaced by wave after wave of immigrants — Irish, French Canadians, Greeks, Poles, Italians, Swedes, Portuguese, Armenians, Lithuanians, Russian Jews, Syrians — most arriving at Ellis Island and passing the Statue of Liberty in search of an American dream. The steerage passengers (3rd class) were registered, physically examined, and they read the mill advertisements, boarded the trains and quickly found work and housing. Each of the immigrant groups would live in separate quarters of a town, striving to break into the social structures and hierarchies of the mills, themselves reflected in make-up of the towns. As one wave of workers moved or was forced out after making wage demands, the oldest remaining (ethnic) group of workers would move into the better quarters, recently vacated, and newest (ethnic) group would take up the recently vacated slums and so on. Together with the ding-dong of the bell, this

hierarchy of living quarters, reflected in the Little Italies, Irish Acres, Frenchtowns and other communities, was the way of life during the second industrial revolution.

EARLY & LATE FORDISM

From the 1750s into the 1920s in England, France, Germany, the United States and elsewhere, textile and other manufacturing mills rose along rivers, newly dug canals and later in urban centres. Once Luddism had been broken in the early nineteenth century, old school or 'old paradigm' weavers and cottage workers – the craftsmen – often had little choice but to join the new cadre of factory workers and live and work according a disciplined regime, later known as Fordism – the system of mass production, initially called the American system of manufacture and still operated by McDonald's and other economies-of-scale producers, that came to dominate industry after industry all over the world.

Fordism is based on at least four principles:

(One) standardised product: not only the product, but each part of the product was standardised, making parts interchangeable. The forerunner was the armaments industry; the American Civil War was fought with mass-produced product.

(Two) worker task routinisation: the worker would perform the same task over and again. It was more efficient, as Adam Smith had pointed out with his famous example of how best to make a pin – in parts, not in wholes. If tasks are performed again and again, they can be mechanised and machines built for single purposes. Like the worker it replaced, the machine could perform only one task. And like the parts the machine produced, the worker was interchangeable.

(Three) worker task efficiency: scientific management specialists, working on the basis of W.F. Taylor's time and motion studies, planned or 'designed' the workers' movements to fit mass production. The stopwatch was their tool, and the flow chart their record. This Taylorist spirit of efficiency had ramifications across society, for example in the rise of an appreciation for behaviouralism, as in the work of B.F. Skinner. The ever quotable Henry Ford said he would have preferred a trained gorilla over the standard factory worker.

(Four) flowline production: first used in the meat-packing industry, the assembly line was perfected by Henry Ford. Product flowed past the workers who no longer moved to and from the product.**(03)**

The system of mass production achieved its aims: economies of scale. After a certain volume of production is reached it becomes progressively cheaper to produce each additional unit, but in order to work, mass production requires mass consumption. Ford understood this and paid his workers well in return for their discipline — that Faustian bargain. With rises in disposable income, new systems of buying on credit and advertising promoting the modernist aesthetic, unprecedented levels of mass consumption were reached by the 1950s.

W.F. Allen, writing about the birth of new-style advertising in the 1920s, explained a dynamic behind mass consumption:**(04)**

> [In 1927 there was] a subtle change of technic. The copywriter was learning to pay less attention to the special qualities and advantages of his product, and more to the study of what the mass of unregenerate mankind wanted – to be young and desirable, to be rich, to keep up with the Joneses, to be envied. The winning method was to associate his product with one or more of these ends, logically or illogically, truthfully or cynically; to draw a lesson from the dramatic case of some imaginary man or woman whose fate was altered by the use of X's soap, to show that in the most fashionable circles people were choosing the right cigarette in blind-fold tests, or to suggest by means of glowing testimonials – often bought and paid for – that the advertised product was used by women of fashion, movie stars, and non-stop flyers. One queen of the films was said to have journeyed from California all the way to New York to spend a single exhausting day being photographed for testimonial purposes in dozens of costumes and using dozens of commercial articles, many of which she had presumably never laid eyes on before – and all because the appearance of these testimonials would help advertise her newest picture. Of what value were sober facts from the laboratory: did not a tooth-powder manufacturer

try to meet the hokum of emotional toothpaste advertising by citing medical authorities, and was not his counter-campaign as a breath in a gale? At the beginning of the decade, advertising had been considered a business... [B]ut by the end of the decade many of its practitioners, observing the overwhelming victory of methods taken over from tabloid journalism, were beginning to refer to it – among themselves – as a racket.

For the consumer planned obsolescence was one outcome of Fordism – durable products did not fit with the cycle of production and consumption. But the pinnacle of Fordism was its Soviet incarnation: the standardised Soviet worker that the socialist planners had in mind would have been the envy of Henry Ford; there was once a barber shop in Moscow with over 100 chairs, turning out more or less identical haircuts. Now there are sculpture pastures with hundreds of identical statues.

(01)
The Unabomber Manifesto is reviewed in Rogers, R., "The Unabomber as Student of Science and Technology?", *EASST REVIEW*, 14, 3, September, 1995, pp. 8-10.

(02)
Murray, R., "Life After Henry (Ford)", *MARXISM TODAY*, October, 1988. The four principles of Fordism and the Moscow barber shop story rely on Murray.

(03)
On Fordism see Chandler, A.D., *THE VISIBLE HAND*, Belknap Press, Cambridge, Mass., 1978. On the clash between mass and craft production, and the survival and reinvention of craft production (or flexible specialisation) especially in regionalist production centres, see Piore, M. & C. Sabel, *THE SECOND INDUSTRIAL DIVIDE*, Free Press, New York, 1984.

(04)
Allen, W.F., *ONLY YESTERDAY*, Bantam, New York, 1957 (1931), pp. 121-122.

03
THE AGE OF SPEED

POPULAR fantasies, dreams and predictions about future technologies have been with us for centuries. As George Basalla points out in THE EVOLUTION OF TECHNOLOGY (1988), the thirteenth century philosopher Roger Bacon felt large self-propelled ships would some day sail the oceans, Leonardo da Vinci prophesied a helicopter and a parachute, Jules Verne the submarine and space ships, H.G. Wells the time machine, Karl Capek the robot and many others much more.**(01)**

There was, however, no imaginative prophecy foreshadowing or driving the development of the railway. No one dreamt of the railway, it came from the other side of light; its origins were in the exploration and exploitation of the underground. The railway shares its roots with the subway, the mechanical lift (and, from an evolutionary perspective, elevators), the steam engine (and engines), air-conditioning (and climate control), and daytime artificial lighting. All came from mining.**(02)**

As with mining and archaeology, which both met with resistance from the Church for penetrating ground (like early surgery for penetrating the body), railways met with resistance of a variety of kinds, mostly that the railway was not natural, and not Godly. It was speedier than nature's horse; it ran on God's Sabbath; it tore up the countryside; it disturbed the pastoral ideal. The garden was disrupted by the machine.**(03)** Like the first anti-machine movement (Luddism), the first environmentalist movement (in the Lakes District of mid nineteenth-century Britain) was directed against technology – the train in relief against the landscape.

The immediate origins and impacts of the train are often invoked when a new technology is introduced and resisted: "they also once thought standing beside a moving train would take your breath away," or such like.**(04)** But the technological culture of the train and its accompanying infrastructural technologies – the schedule and the telegraph – are less readily applied because their impact is much more abstracted. When thinking about the technological culture in which we live, or the culture that grows around new technologies like the mobile telephone, it's important to recall that abstraction and draw parallels with the pervasive technological culture of the 'factory town' and then the 'railway nation'.

The bells of nineteenth century factory towns regulated the lives of the local inhabitants. Whenever the bells were rung, usually in unison, the town's inhabitants would refer to the omnipresent clock tower, working and living according to the dictated rhythm. Their time no longer organised and experienced naturally, but artificially. Factory time was not farmer's time. For many, the mechanical clock and bell had replaced the sun as life's regulator. Significantly, the inhabitants of factory towns had little need to keep their own time or make their own schedules as these were were kept for them. Bell-ringing schedules were printed and distributed, and eventually made open to negotiation between management and unions, but otherwise these schedules did not need to be referred to very often. At home or in the factory, the worker (and everybody else in town) knew what each toll of the bell signified.

A FAST AND FURIOUS GAME FEATURING FAMOUS BRITISH SPEED MACHINES OF PAST & PRESENT. CARS, PLANES, SHIPS, TRAINS - TIMELESS CLASSICS ALL INDIVIDUALLY ILLUSTRATED. FOR PLAYERS 6 YEARS OLD UPWARDS.

While more frequent and totalitarian, bell-ringing in factory towns was somewhat similar to that in towns dominated by the Church. It was similar in the sense that the tolling of the bell, and the system of discipline behind it, extended, in the first instance, only as far as earshot. Time, while based on the potentially universal or 'objective' sources as the sun and the stars, and embodied in mechanical clocks beginning in the thirteenth century, was experienced locally.

The sense that time is local and a matter of local experience, dictated by the sun and the stars (God) or by church bells (still God) or factory bells (Capital), changed dramatically in the late nineteenth century thanks to the efforts of the railways in following the paradigm of standardisation. As Stephen Kern points out in his book THE CULTURE OF TIME AND SPACE (1983), on a cross-country train trip in the United States, which was possible from the late 1860s, the passenger would pass through some two hundred different local time areas. The situation was similarly 'chaotic' on other long-distance routes, like the Trans-Siberian Express from Moscow to Vladivostok or the early Orient Express, from Paris to Varne on the Danube, and eventually to Sofia and Constantinople.

In the nineteenth century time differences were not measured in single hours but in hours and minutes. Cities or regions would have their own times based on solar or astronomical readings. New York was five minutes behind Philadelphia; Paris nine minutes and twenty-one seconds ahead of London. With the coming of long-distance trains, the railways desired to standardise time to coordinate their schedules from single hubs. If time were standardised there could be one schedule, read from the top down or from the bottom up. The railways thought it was simpler, more logical and more practical to have one set of times.

The railways' drive to standardise time resulted in a major international conference in Washington DC in 1884, whereby twenty five 'railway nations' proposed Greenwich, with its time-keeping observatory, as the zero meridian. Time would begin in London and the rest of the world would be carved up into twenty four zones, each zone representing one hour. An artificial international date line was also set, changing the day of the week when crossed. It is no coincidence that the anarchist in Joseph Conrad's THE SECRET AGENT (1907) wished to destroy the Greenwich Observatory and thus World Standard Time — the regulation of life by a single centre.

In the years after the Prime Meridian Conference the railway nations — including Japan, the Netherlands, Belgium, Germany, Austro-Hungary, Italy — signed World Standard Time into national law. Of the railway nations, the French delayed its adoption the longest, finally agreeing to Greenwich Mean Time in 1912, in a diplomatic bargain at the World Conference on Time held in Paris. Paris time would be turned back by nine minutes and twenty one seconds, and broadcast to Algeria and beyond by the wireless from the Eiffel Tower. Eventually, national time broadcasts would shift to national radio stations like the BBC World Service, whose hourly rhythm hasn't changed for decades.

In 1916, AT&T requested the French government for use of the Eiffel Tower for the first transatlantic wireless experiments, approved for only one hour in the middle of the night. War allowed only limited experimentation since troop and supply movements were being relayed via the Eiffel Tower – the highest available point. The promise, in Western Europe, that standardisation of time would bring peace hadn't panned out. Russia was one of the few countries that recognised early on the potential impact of railways in wartime. Unlike most of Europe, Spain and Portugal being the other exceptions, Russia decided to build railways with broader gauges so that trains from Europe, possibly with troops and supplies on board, could not pass its border, although this caused commercial bottlenecks in peacetime. An analogous situation is now found on the border between Iran and Turkmenistan.

With the standardisation of the railway gauge, the railway timetable and world time also came the proliferation of the pocket watch, which by the turn-of-the-century was being mass produced. Owned by the bourgeoisie as well as the lower middle class – office clerks and better paid factory workers like the mechanic in Charlie Chaplin's *MODERN TIMES* – the pocket watch had replaced the bells as man's timekeeper. There was one time and watches could be set to it at the railway station, considered to have the most accurate and accessible time. The bells would still ring, but they hardly dictated behaviour. Public time – or Church time or factory time and then railway time – had become private time.

Punctuality, or public time obedience, took longer to manifest itself as a dominant cultural value in the west.**(05)** It diffused not only from factory and railway regimes, but also from the army and private (boarding) schools which made punctuality an internal law, variously punishable. Intriguingly, the French national railways took into account the difficulties of shifting to a standardised railway time; two times were kept at the stations, one in the waiting rooms and one on the platforms. The waiting room clocks were five minutes ahead of those on the platforms, allowing the passengers with a slower, natural or fashionably late private rhythm, to board the trains 'on [public] time'.**(06)**

The widespread acceptance of the standard measurement of time in relation first to factory work then to travel was of primary importance in the development of an age of speed. The quest for speed was a product of Fordist-style capitalism, moving goods to market for quicker consumption, as mass production required mass consumption at a pace proportional to production and distribution. This dynamic underlaid managing time as never before, but there is also a different argument to do with the world's records culture, national pride, national competition and newspaper reporting – other seminal reasons behind what William Blake, in a different period, called the 'annihilation' of time and space.

Behind the quest for speed has been competition within and between modes of transport. In Britain and the United States, competition between railways was institutionalised from the beginning. This was not the case in continental Europe which preferred regulated monopoly. In Great Britain there were two railways offering service from Dover to London – and, by extension from Paris to London. Competition was fierce, pricing and advertising campaigns were fought for new business. Speed was also an issue.

In the early to mid nineteenth-century newspapers would hire boats and trains to race the news from Paris for the latest edition of the paper, paying handsomely the fastest boats and the fastest trains. Speed, the competition between modes of transport and between newspapers for the latest news, was behind the latest story. Gradually, however, speed became the story and newspapers, already accustomed to their own speed competitions, would offer prizes for ships, cars and planes to break newly created world speed records. Governments followed suit and offered prizes as well. Early aeroplane advances are attributable in part to the financial gain and press prestige won by the victors of these races. While the telegraph had ended the exciting part of the race for the news, reporters were still racing to get to the wire first with their stories.

An important shift, from the national to the international, took place during the Age of Empire, as the historian Eric Hobsbawn calls it, underpinned by nationalism. Prior to the turn of the century, competition was shifting from modes of transport within a country to modes of transport between countries — from intra to inter, national to international. National competition, out of wartime, also occured in the context of the World's Fairs beginning in 1851 and the modern Olympics from 1896. The race between nationals (and by extension between nations) was the story — the race across the English Channel, to the source of the Nile, across the Atlantic Ocean, to the North Pole, to the South Pole, and around the world. The publication of Jules Verne's *AROUND THE WORLD IN EIGHTY DAYS*, in 1873, perhaps marks the symbolic beginning of the 'race' and the interest of the newspapers in sponsoring and writing about the race. Eighty days was the time to beat, and there'd be sponsors lining up and prizes being devised to promote and cover any such race in future.

Nowadays that speed culture has become quaint. When the Channel Tunnel opened a few of years ago, *THE TIMES* of London, characteristically, held a contest between modes of transport from London to Paris. Reporters boarded the tunnel train, a car (with ferry link) and the airplane at about at the same time, with the Arc de Triomphe as their destination. The airborne reporter won. It was front-page news, albeit in the style of the *GUINNESS BOOK OF WORLD RECORDS* (first published in 1955), the great trivialiser of records. For opponents of the speed culture (or sustainability proponents), it would be easy to demonstrate that there is hardly a grander destination these days than the Guinness book.

(01)
The submarine is in part the product of the American Civil War. The most interesting relationship between science fiction writing and technology development most likely will be found in the history of the Soviet technology, and the enduring significance of space travel [and Mars] in Russia. For a western perspective on this relationship, see the series of articles by I.F. Clarke in the journal, *FUTURES*, in 1990 and 1991.

(02)
See Williams, R., *NOTES ON THE UNDERGROUND*, MIT Press, Cambridge, Mass., 1990.

(03)
For Emerson's and Thoreau's reactions to the train, see Marx, L., *THE MACHINE IN THE GARDEN*, Oxford Univ. Press, New York, 1964. See also Faith, N., *THE WORLD THE RAILWAYS MADE*, Pimlico, London, 1994.

(04)
Joerges, B., "Expertise Lost: An Early Case of Technology Assessment", *SOCIAL STUDIES OF SCIENCE*, 24, 1994, pp. 96-104.

(05)
Pursell, C., *WHITE HEAT: PEOPLE AND TECHNOLOGY*, University of California Press, 1994, chapter 3.

(06)
Kern, S., *THE CULTURE OF TIME AND SPACE*, Weidenfeld & Nicolson, London, 1983, p. 13.

04
MAKING USABLE HISTORY FROM SYSTEMS BATTLES

THERE are no definitive histories, only those currently considered more plausible than others. In the history of technology and culture, four modes of explanation of social and technical change have long been considered plausible – the progressivist-materialist, revolutionary, hagiographic and evolutionary perspectives. These modes, however, contain certain contextual biases. We should be aware of the biases for at least two reasons: first, to avoid simply incorporating ideas about social and technological change derived from other historical periods into our own projects, and second, because the identification of a bias is the first step in 'debunking' or undermining someone else's arguments and improving one's own.

Change has long been synonymous with advance or progress, especially technologically. This mode of explanation is the dominant bias, 'pioneered' by archaeologists. In a sense archaeology bequeathed the idea that technical advance is synonymous with social or civilisational advance, as its practitioners dubbed historical periods according to the materials used to make tools (technological artefacts), i.e. the stone age, bronze age and iron age.**(01)** It's much easier to name past periods after the material remains found buried in the ground than after incomplete data on 'social relations' or the 'division of labour'. But this tradition at least connotes that change came about through the 'discovery' and application of materials. Archaeology, a scientific profession which, like geology, burst onto the scene in the nineteenth century, has had a 'material' bias, a bias reflected in the expression, 'to bomb a country back into the stone age', implying that by destroying material-based tools and technologies the process of civilisation will have to begin all over again.

The second mode of explanation of change rests on the idea of revolution, an idea inherited from history-writing about the eighteenth century – the age of glorious revolution, in France. The idea that technological change is a) sudden and b) brings about a concomitant social revolution comes upon us in our newspapers, as in an *INTERNATIONAL HERALD TRIBUNE* headline of a few years ago, "The 747 Jumbo Jet: 25 Years of Social Revolution". Here it seems that the jet itself caused a social revolution.

The third mode of explanation of change comes from the late eighteenth century, often dubbed the 'age of genius'. Some time ago, newspapers ran a story claiming that listening to Mozart increases levels of creativity because the patterns in his musical compositions correlate with the most active human brain waves, which certain scanners can measure. 'Genius' has been attributed not only to those like Mozart and Beethoven but also to such figures as Thomas Alva Edison, the 'wizard of Menlo Park'. Edison may have enjoyed and exploited the moniker, but he did attempt to debunk notions of genius as the source of inventive activity with well-known slogans like: "Invention is 1% inspiration, and 99% perspiration." However, the 'creative genius' bias is a familiar trope appropriated by biographers of great inventors. Edison's most famous invention, the electric lightbulb, has become synonymous with the sudden birth of an idea – a flash of brilliance, inspiration.

Evolutionary change, the fourth mode of explanation, stands in sharp contrast to the revolutionary but still shares the notion of progress. The Darwinian and biological origins and the connotations of evolution are well understood, and its use as a mode of explanation for technological change can be attributed to museums. Early evolutionary descriptions of technical change were used by industrial museums, which based their methods of presentation on those of earlier natural history museums. As the evolution of man, so the evolution of technology. From ape to homo erectus to modern-day human; from Daedulus' wings to the Wright Brothers to Boeing's 747s.**(02)**

Over the past twenty years history-writing on technology has suffered from these perspectives because the authors feel that their readership are 'convinced' that the biases are true. Most books on the history of technology and culture consume too much space arguing that technological change doesn't simply imply social change, that technological advance does not imply social or civilisational advance, that new technology is not the product of an Edisonian flash of genius and that great men have not been the source of most inventions, often times not even the 'greatest' inventions. The biases have become 'bogeys' to be slain.

In the SCIENCE IN AMERICAN LIFE exhibition at the Smithsonian American History Museum, the display of great African-American female inventors and scientists falls into the politically correct, bogey-fighting trap — reinforcing the biases by employing the same terms of reference — by replacing great men with great women and African-Americans. One should learn about the bogeys, how to avoid them, but also — and this is more difficult — to avoid writing about them. In some sense, a bogey exists relative to the amount of space devoted to slaying it. In the great truth-seeking, bias-debunking tradition of academic scholarship, there will then rise a group of writers arguing that the men were indeed great. Little advance of thought will be made.

SYSTEMS BATTLES

The rationale for looking closely into the early history of current dominant systems relates not only to arguing for a fifth mode of explanation for technological change, but also to challenging the commonplace idea (or 'received truth') that the marketplace sorts out the 'best' technology and that the consumer and society are the beneficiaries.

One of the notable challengers of received truth was Robert Fogel, a 'cliometrician' or quantitative economic historian awarded a Nobel Prize in economics. Fogel's work, beginning with RAILROADS AND AMERICAN ECONOMIC GROWTH (1964), was dubbed 'counter-factual', and helped to lay the groundwork for the conditional or 'if/then' perspective in both historical scholarship and in (science fiction) literature, as Bruce Sterling and William Gibson's THE DIFFERENCE ENGINE that asks the question, what if Charles Babbage's protocomputer had been built? Contrary to popular opinion (an expression employed, strategically, to construct and set up a bias ripe for debunking), Fogel argued that the coming of the railway was not 'indispensable' for economic growth in mid to late nineteenth-century

America. On the contrary, the existing canal system would have been sufficient for similar levels of growth. Though he didn't define it as such, he was writing about a systems battle, and challenging a dominant mode of explanation about the effects of the victor. Instead of the 'effects' of the triumphant system, the 'alternative paths' or 'roads-not-taken' historians examine the effects on society (and increasingly the environment) of having lost a potentially viable system – technology opportunity cost.

Recently historians have gone back to the early history of cars, driven by the internal combustion engine, in order to assess whether that engine was indeed the best, or whether other forces were at work in the triumph of the 'big polluter' over the steam car and the electric car. Edison, not surprisingly, was pushing for the electric car, and before Ford started mass producing the Model T in 1913 and exporting 'Fordism' to Hitler and Stalin's production regimes, the electric car, according to the historians, had excellent prospects.**(03)** Another recent contender to replace the petrol car, at least for daily travel to work, has been teleworking through the Internet. In a mid-Nineties study, Wolfgang Sachs of the Wuppertal Institute learned that the fabrication of one computer requires fifteen to nineteen tons of energy and materials, compared to twenty five tons for the manufacture of an average car.**(04)** For the environment, the substitution of personal car for teleworking wouldn't be that revolutionary after all.

A number of systems battles point more readily to 'stasis' as opposed to great change, and reveal more generally the conservatism of technological change. New technology was often not thought of as revolutionary; in the first instance new technologies often reflect and reify, rather than change, social relations.

The conservatism of technological change is illustrated by the following story from the early history of the telephone.**(05)** Alexander Graham Bell's 1876 patent application for his telephone read 'improvement of telegraphy', that is the wired telegraph, using Morse code, put into service by Samuel Morse, in Baltimore with the early railways, some forty years earlier. Bell wanted to sell his patent for the telephone immediately after he displayed at it the World's Fair – the Centennial – in Philadelphia in 1876. He offered it to Western Union, the telegraph giant, in negotiations of 1876 and 1877 for $100,000, but they refused, thinking that the direct communication afforded by the 'intruding' democratic telephone – democratic not in its price-tag but in the fact that it required no training to use – did not fit with the hierarchical manner in which business was done at the time – written messages, transmitted in code on the telegraph, decoded and written out by the skilled operator and brought by messengers. (Machine-written correspondence on the typewriter was also not taken to enthusiastically for years.)

Shortly thereafter, Western Union changed its mind and started buying more technically advanced telephone equipment from Elisha Gray and Thomas Edison. Now both the Bell Company and Western Union had set up communication exchanges in cities, offering both telephone and telegraph services (the telegraph patent had expired). Competition was fierce.

Bell sued Western Union for patent infringement on the telephone, and negotiations between the two companies commenced. Western Union agreed to settle out of court (much to the delight of a smaller and financially weak Bell Company) as long as Bell agreed to keep to the telephone business and allow Western Union to dominate the telegraph business. Western Union felt that the lucrative business market wouldn't switch to the telephone, whereby subordinates (it was thought) could ring up their superiors directly. Western Union felt the business community was not ready for technological change. In the United States, the last Western Union telegram was delivered years ago.

Belief in the staying power of the telegraph illustrates the conservatism with which technological change is met, and how the use of new technologies follows not necessarily from their functionality, but from the technological culture of use in which they are introduced. In France, initially the telephone was not allowed to be used for two-way private communication, but one-way and only by the military.

Other popular explanations of how a battle of the systems may be lost through a lack of 'foresight' include conspiracy of the powerful, or the interest model. Two well-known stories will illustrate this; one concerns dominant industry resistance to technological change by those who stand to lose, the other illustrates how conservatism is defeated by those who stand to gain.

FM radio (Frequency Modulation) was patented by Major Edwin Armstrong in the early 1930s, about a decade after the formation of RCA, the Radio Corporation of America. RCA was set up by the US government, in a patent-pooling agreement in the 1920s between AT&T, Westinghouse, and General Electric to compete against the British Marconi Company, which was making inroads in the fledging US markets. RCA broadcasted on AM, amplitude modification, at lower frequencies. Early promoters of FM thought it couldn't lose because it was technologically superior; FM produced no static and high fidelity sound at higher frequencies. Armstrong had a small but decent market for his invention, pioneered in New Jersey, by the early 1940s.

With the advent of the television, just after World War II the FCC, the U.S. authority assigning portions of the electromagnetic frequency spectrum, made a crucial decision to keep AM where it was, move a television channel into the old FM frequency and move FM 'upstairs' to a higher frequency. This made all the FM radios and transmitting equipment obsolete overnight. A dejected Armstrong felt the decision was the result of the conspiracy of powerful 'hidden forces' operating to keep the technically inferior AM radio as the system of choice. In the event, AM remained dominant well into the 1970s, as testified by American stock car radios from the period. In 1954 "twenty-three years to the day after patenting FM, Major Armstrong (born of a devoutly Presbyterian family in Manhattan) put on his hat, coat, scarf and gloves, and walked out his apartment window, thirteen floors to his death," as one may read on the Internet. Death by dominant industry resistance to change. Death by lack of acceptance of a technological revolution, and a lost systems battle.**(06)**

The second story of an interest-driven conspiracy of the powerful concerns the rise of bus technology. Just after World War I one of the largest electric streetcar systems in the world was located in Los Angeles, California. The neighbourhoods in the dispersed city grew up along the streetcar lines, as the trolley companies bought large tracts of land, built houses and laid track to the new suburbs (and elsewhere to the new amusement parks, often owned by the trolley companies). As the Californian market for automobiles was growing at a rate larger than anywhere in the world, General Motors, together with Standard Oil, formed a subsidiary of National City Lines to purchase the city trolley lines. Beginning in the early 1940s they tore up the track and replaced the streetcars with diesel bus lines. Across the country they repeated the strategy that had been so effective in California. An investigation in the practices of National City Lines was undertaken by the U.S. Senate in 1974, and the conspiracy was documented, and later debunked.**(07)** Whether fully orchestrated by automotive interests, the demise of convivial public transportation followed.

Other well-known battles include that between the gas and electric refrigerator and between AC and DC electricity. Both hinged on campaigns directed at the safety of an older system compared to a newer system. The defeat of the gas fridge in favour of the electric fridge had much to do with the public relations campaign waged by General Electric that gas was risky. The twenty-four-hour-running electric fridge, along with many other electrical appliances, would also enable the electricity companies to increase usage during off-peak hours. (Incidentally, some of the new 'green fridges' are gas fridges.) Edison attempted a similar public relations ploy to promote his Direct Current (DC) over Westinghouse's Alternating Current (AC) system in the late 1880s and early 1890s. Among other stratagems, he engineered the New York State Legislature to change their technology of execution from hanging to death by electric chair. Edison had his engineers orchestrate the installation of an AC system for the first electric chairs, and subsequently advertised that the use of AC for electrocution proved it to be the more dangerous system.

Paul David, the roads-not-taken economist, has written about AC/DC battle as well as the staying power of the QWERTY keyboard over the DVORAK keyboard, patented in 1932 by August Dvorak.**(08)** In the nineteenth century the QWERTY keyboard was designed to slow down typing speeds so typists wouldn't jam the machine. With DVORAK typists have broken world typing speed records, one measure of the 'better' system. Another would be the repetitive stress injuries associated with QWERTY.

When new and 'better' technological systems are trumpeted, it is worth recalling these and other specific examples of lost battles, from the levels of abstraction of craft versus mass production down to that of keyboard layouts. In confronting better technologies of the future, the question always remains 'better for whom'?

To conclude, here is a basic idea about how to supplant an existing technological system which has become so hard, so obdurate, so entrenched in society, that it seems impossible to dislodge. Not only roads-not-taken historians but contemporary sociologists of technology

...ard Layout

&	*	(	)	[	+	Del
	8	9	0	]	=	

G	C	R	L	? /	{ }	| \\

H	T	N	S	_ -	Return

B	M	W	V	Z	Shift

⌘	Option	Ctrl

are endeavouring to take on the hardest case: the car. The challenge the automobile poses is formidable, largely because it forms part of an entrenched technological system encompassing roads, gas stations, oil companies, automobile retailers, repair shops and car drivers. Somewhere around 1973, corresponding with the oil crisis and the rise of a kind of environmentalism which challenged natural resource and space consumption, the car became a serious concern. To address this concern the U.S. introduced gasohol and the catalytic converter, while France (for example) developed the TGV. Gasohol and catalytic converters are technical fixes allowing a system to remain in place, the TGV, however, challenges ideas of mobility, of how to get from A to B. It is often observed that for a new technological system to replace an existing, entrenched system, new cultural concepts (of mobility) have to be introduced and practiced in social experiments — in this case, less spontaneous mobility, less individual control and volition, more relaxation during transportation.

There is a great deal of research into strategies to move beyond automobilismus. Two generic strategies, technology forcing and strategic niche development, are compared.**(09)** An example of technological forcing was Californian legislation directed at car manufacturers: to continue to operate in the state, a percentage of annual car sales would have to be electric. Examples of strategic niche development — research and development space providing short-term protection from the market — would be trajectory-breaking programmes sponsored by government, certain research arms of larger corporations, or even specific companies, such as Interval Research Corporation in Palo Alto, conceived to provide fertile ground for projects which should spin off once they are ripe for the market. Time is given to develop ideas and products which in the first instance do not have a clear-cut market.

It has been argued that for a new technological system to break through, a combination of the two should be pursued. Technology is developed in a protected space, preferably together with a core user group and a group of regulators. The government support the project by inviting industries to pursue it, hinting at regulatory penalty (i.e., the impending forcing of a technology through legislation).

An important part of convincing all to participate is to argue that the project has a future, in fact is the future. If, however, those futuristic arguments are biased towards materialism, revolution, individual genius or vulgar evolution, they are ripe for debunking.

(01)
See Mumford, L., *TECHNICS AND CIVILIZATION*, Routledge, London, 1946 (1934). See also the work by V. Gordon Childe.

(02)
Rogers, R., "Exposities van technologie voorbij 'goed' of 'kwaad'", *ZENO*, 4, September, 1994, pp. 14-17.

(03)
On the early history of steam and electric cars, see Schiffer, M., *TAKING CHARGE: THE ELECTRIC AUTOMOBILE IN AMERICA*, Smithsonian, Washington, DC, 1994. On Ford, see Flink, J., *THE AUTOMOBILE AGE*, MIT Press, Cambridge, Mass., 1988.

(04)
Sachs, W., I. Illich, S. George and A. Ross, "Slow is Beautiful", *NEW PERSPECTIVES QUARTERLY*, 14, 1, Winter, 1997, pp. 4-17, reprinted in Millar, J. and M. Schwarz (eds.), *SPEED – VISIONS OF AN ACCELERATED AGE*, The Photographers' Gallery, London, 1998. Key research on teleworking and personal mobility is summarised in Mokhtarian, P., "Now That Travel can be Virtual, Will Congestion Virtually Disappear?", *SCIENTIFIC AMERICAN*, October, 1997, p. 61.

(05)
Rogers, R., "Visions Dancing in Engineers' Heads: AT&T's Quest for a Universal Telephone System", WZB Discussion Paper, FS II 90-102, Science Centre Berlin, 1990.

(06)
On Armstrong, see the October 1996 edition of the journal *TECHNOLOGY & CULTURE*.

(07)
See Jim Klein and Martha Olson's documentary, *TAKEN FOR A RIDE*, a co-presentation by the Corporation for Public Broadcasting and the Independent Television Service, USA, 1996. The conspiracy theory is debunked in Adler, S., "The Transformation of the Pacific Electric Railway", *URBAN AFFAIRS QUARTERLY*, 27, 1, September, 1991, pp. 51-86.

(08)
David, P., "Clio and the Economics of QWERTY", *AMERICAN ECONOMIC REVIEW*, 75, 2, May 1985, pp. 332-337.

(09)
Schot, J., R. Hoogma and B. Elzen, "Strategies for Shifting Technological Systems. The case of the automobile system", *FUTURES*, 26, 1994, pp. 1060-1076. See also Weber, M., R. Hoogma, B. Lane, J. Schot et al., *EXPANDING TECHNOLOGICAL NICHES. HOW TO MANAGE EXPERIMENTS WITH SUSTAINABLE TRANSPORT TECHNOLOGIES*, University of Twente, 1999.

05

SUBLIME CONTROL: LEGACIES OF THE ATOMIC & ANTI-ATOMIC AGES

As the seconds were counted down, the world listened in silence to the live radio broadcast. **(01)**

And thus the natives express to the people of the United States their welcome, despite the fact that the Atoll Bikini may be utterly destroyed come July the 1st. But the natives, in their simplicity, and their pleasantness and their courtesy are more than willing to cooperate, although they don't understand the world of nuclear energy anymore than we do.

American officials discuss plans to prepare the natives for evacuation of the Atoll. The islanders are a nomadic group, and are well-pleased that the Yanks are going to add a little variety to their lives.

And here, by the way, you hear them singing their Marshallese version of 'You Are My Sunshine'.
Firing time.
And I have just been informed... Thirty seconds... that is the voice at the controls desk, marking the seconds remaining.
The final switches have been thrown. No one can stop it.
The atomic bomb is about to explode.
Firing Time.
Twenty seconds.
We do not know how it's going to sound, but 42,000 men here are watching...
All the observer ships... Ten seconds... All of the observer ships are in position in the open sea. They are about 10 miles away.
Firing time.
Five, four, three, two, one.......... **(02)**

THE TECHNOLOGICAL SUBLIME

The beautiful and the terrifying — that which we do not control. Awe-inspiring and fearsome — beyond our control because it's beyond our ken, beyond our realm of experience and beyond our ability to assess its impact in advance. It is to be experienced first; understood only later, if ever. As certain 'veterans of the sublime experience' of the first atomic bomb blast put it: you really had to be there.

TECHNOLOGICAL LANDSCAPES

The explosion of the first atomic bomb ('Fat Man') — at the secret Trinity Test Site in Alamagordo, New Mexico on July 16th 1945 — has been called the single most important event in the history of the twentieth century. Its legacies in western technological culture endure. Four hours after the successful test, a ship left San Francisco bound for the American air base at Tinian Island in the South Pacific. It was carrying 'Little Man' — the bomb that would be dropped on Hiroshima twenty-one days later (the ship was torpedoed by a Japanese submarine on its return trip killing five hundred crew members). Departing from the Tinian air base the Enola Gay, a Boeing B-29 from Seattle, delivered its cargo on August 6th. Its mushroom cloud would symbolise the nuclear age and the spoils of the war, Pacific Islands either captured or claimed, would be the initial testing grounds.

While radioactivity was 'discovered' in 1896 and nuclear fission in 1940, the detonations at Trinity, Hiroshima and three days later at Nagasaki 'opened the circle' to military and civilian applications of nuclear power that the Comprehensive Test Ban Treaty and the Department of Energy's planned permanent waste sites in New Mexico and Nevada are trying to close.(03)

The United States was the first of the big five nuclear nations, all of whom sit on the UN Security Council — the first American (or Allied Anglo-American) test was followed by those of the Soviet Union (1949), United Kingdom (1952), France (1960), China (1964) — and thereafter India (1974), Israel (1979). More recently Pakistan joined the ranks, though appear not to have detonated a bomb. South Africa has built, but not detonated, seven bombs; Argentina and Brazil decided against making them. Who will be next? When we learn of a country approaching the list of atom bomb nations, we may not only 'remember' the Cold War of bomb shelters, scares, mad scientists and Ronald Reagan's "five minutes till bombing", or the Union of Concerned Scientists' *DOOMSDAY CLOCK*, that once reached two minutes to twelve, but also the question posed when confronting a new technological system, perhaps the most significant legacy of the atomic age. Despite the endless promise of a better world of tomorrow, what regime of control does the new system imply? The answers to this question for any technology are informed by the tangible lessons of the atomic age, still very much with us — especially when we learn that protesters know when the next shipment of radioactive materials will take place.

The first UN resolution, passed by the General Assembly in 1946, called for the elimination nuclear weapons. The U.S., for example, produced nuclear weapons through 1989. While there's talk about the need to perfect 'simulation tests', the French nuclear tests off Mururoa and the Chinese tests shortly thereafter were performed with new weapon systems in mind. Though the laboratories in the three original secret cities in the U.S. — Hanford, Washington; Los Alamos, New Mexico; and Oak Ridge, Tennessee — have been scaled back, they will not be closed for fear of lost know-how. The U.S. Secretary of Defense recently stated that, "you can't mothball intellectual capital". To him, know-how is not historically cumulative, to be pulled off the shelf in times of need. For others the crucial argument for keeping 'nuclear capability' and stockpile is simply that the 'genie's out of the bottle'. Failing the destruction of the world, through natural or man-made disaster, including, paradoxically, an all-out nuclear war, the genie cannot be rebottled. In this sense, there is a fear that bomb-making knowledge is here to stay. The nuclear age rests on a number of these seeming contradictions. Indeed, the U.S. government has considered the contingency of world destruction, and working with linguists and futurologists, is devising warning barriers and signs in pictograms and in a variety of languages next to the permanent waste site in New Mexico.

In an international relations course at Cornell University, Professor Richard Ned Lebow once came dressed as Dr. Strangelove, Stanley Kubrik's Cold War scientist film character (1962) who couldn't control his arm.**(04)** He had to press the button, the scientist's creation had to be experienced. Lebow, who'd worked in the CIA's psychology department during the Carter administration, explained that Mutual Assured Destruction (MAD) — the Anglo-American policy that deterred the superpowers from pressing the button and unleashing nuclear war — was actually based on the image of a Dr. Strangelove. For MAD to work, the superpower leaders had to and did indeed convince each other that they were sufficiently mad to press the button. If they were to act 'rationally', the button would never be pushed. The French, who were marginally involved in the Manhattan Project and left NATO in the late 1960s under de Gaulle, had an independent policy (like the Soviet's) called 'proportional deterrence'. The French communicated to the Soviets (and the Soviets to the west) that if attacked, they retained enough firepower to obliterate a healthy number of cities. Recently, the French offered to provide the European Union's 'nuclear umbrella', separate from NATO. In the meantime, they've rejoined the defence alliance, but retain an independent nuclear capability.

David Nye, in *AMERICAN TECHNOLOGICAL SUBLIME* (1994), sums up the reasons for making and dropping the bomb as "patriotism, the advancement of science, and the protection of the Free World". To that must be added ending World War II, and winning the bomb race initially against the Nazis, then subsequently, but more vaguely, against the Japanese. If not the first military application, then certainly subsequent explosions had the purpose of intimidating the Soviet Union. Once the Soviet Union had exploded their first bomb, the race was for the bigger and bigger bomb — first culminating in the 1954 H-bomb detonation, 'Project BRAVO'. The H-bomb — 15 megatons — vaporised three uninhabited Pacific islands and, after a sudden wind change, sent a shower of radioactive debris miles around, also infecting a boatful of Japanese fishermen. 'No need to worry about the natives' was the message communicated to the American people and to the world. Secretly, the U.S. sent in a monitoring team.

There were other reasons behind the detonation of the bomb, which guide considerations of the legacies of the atomic age. One may be the 'technological imperative', or indeed the sublime technological imperative. In a sense the detonation at Trinity brought the eighteenth-century pursuit of the 'sublime' full circle. Detonating the bomb became an almighty technological experience, met sublimely — first with awe, since it was 'beautifully hypnotic', then immediately thereafter with terror. After the tanning butter was applied and the bomb was detonated, the flash witnessed, the pressure wave felt, and the excitement subsided, words were found to describe the terror. The lead physicist, Robert Oppenheimer, invoked Hindu: "I am become death, shatterer of worlds." Lewis Mumford, a leading intellectual, entitled his article: "Gentlemen, you are mad!"

There was and remains the fascination for experiencing the event. From time to time one will read gender and technology scholars argue that we must avoid the pitfalls of the technological sublime. An article by a *SEATTLE TIMES* journalist, written on the fiftieth anniversary of Trinity, appears to reflect this:

> *White-male stereotypes have been beaten to death and yet Los Alamos is unmistakably, even to this white male, an overwhelmingly 'guy' kind of place: a techno-freak haven for boys who like things that go boom, even underground. The enthusiasm for this raw power can be slightly infectious, even during a brief visit: I found myself musing that once, just once, it would be awfully interesting to stand in the Nevada desert and see one of those things go off.*

The journalist weaves together three elements of the bomb-making culture: an unmistakably male laboratory environment, a yearning for power delivered in the form of an explosion and the infectious expectation of experiencing a sublime moment, which somehow places forethought about the implications of the event on the backburner. More worryingly, David Nye also talks of the atomic scientists' maternal descriptions of the bomb and the names given to them.

Two elements made the fabrication of the bomb itself a compelling experience, devoid in the first instance of moral responsibility. It wasn't only the quest for the sublime moment in a 'guy's world', it wasn't only a fascinating puzzle for those few who knew the ultimate goal, but for that vast majority working on each small piece of the puzzle. The Manhattan Project was split up into task forces working on separate pieces, allowing for the vast majority of scientists to concentrate on their small puzzles according to the principles of scientific bureaucracy, 'characterised by specialisation of functions, adherence to fixed rules, and hierarchy of authority'. As Zygmunt Bauman has written about in reference to the bureaucratic administration of the Holocaust, compartmentalisation of tasks allows individual workers or teams to fix their minds solely on the puzzle at hand. Bureaucratic administration, a variation of Taylorism and Fordism, provides enough distance from the ultimate aim, and thus provided individual scientists enough distance from the morality of the ultimate aim to justify their work, to themselves and to others. Of course, a sense of moral purpose was derived from producing a war-ending device, until the puzzle had been solved and the lead scientists gained a fuller sense of what they'd made. Some entered a betting pool, guessing the might of the explosion. Illustrating the normality of the extraordinary (another phrase often used in connection with the administration of the Holocaust), money changed hands after the event.

The second element relates to normal puzzle-solving in engineering, the drive for the 'technically sweet' solution (the bomb was described as 'technically sweet' by Oppenheimer himself). Others have called this 'engineering virtuosity' — the aesthetic beauty of the technical solution to a theoretical problem, or its beautiful material realisation. Before it was detonated, the bomb, while rather ugly on the outside, was beautiful on the inside. Perhaps it's not difficult to imagine some people and engineers (shall we call them mad?) finding both the outside and the inside of current cruise missiles aesthetically pleasing. When Klaus Fuchs delivered atomic bomb plans to the Soviets in 1949 the literal copying of the plans could be interpreted not just in terms of Soviet 'scientific backwardness', their inability to improve upon them, and their five-year 'deadline', but in terms of the reverence they showed towards the beautiful solution to the problem contrived by their theoretical counterparts — kindred spirits in science. In the event, the Soviets made the same wiring 'mistakes' as the Manhattan Project scientists. (The Soviet proclivity for copying high-technology blueprints — in this case, purposively faked blueprints — is said to have been behind the supersonic air disaster at the Paris air show in 1973.)

Once the bomb puzzle had been solved, and the leading scientists were able to take some distance from their activities, a large percentage of them did not wish their bomb to be deployed. Leo Szilard organised a petition to put "on record... opposition on moral grounds". For those few scientists (less than one hundred) even the sublime explosion didn't have to experienced; the solution, the rendering of the technically sweet, had been reward enough. The petition became more urgent after the Trinity test. Much has been written about the days between July 16th and August 6th 1945 and the atomic scientists' last-minute efforts to convince President Truman and the Secretary of Defense to either keep it secret or only demonstrate it. It could be said that the technological imperative was a major factor behind its use; there was a war to end, and the Soviets to intimidate, too.

Niels Bohr, the physicist, once said to Edward Teller, the inventor of the Hydrogen bomb, "I told you it couldn't be done without turning the whole country into a factory. You have just done that." When confronting decisions about new technological systems — including the normal and the fascinating, potentially sublime-producing systems — one ought to ask the question, what form of social organisation and control does the system (or could the system) imply? Nuclear bomb-making, and by extension, nuclear power could very well imply a strict, centralised, and secretive regime of control.[05] Whether such a regime of control is intrinsic to the technology is a matter for debate: it could be said that it's intrinsic to the context of use, to the context of war-time use, the context of mutual assured destruction, in that only a 'controlled' nuclear capability (with willing destructors) prevents war. The ground for war is prepared once control over the war-making components is lost.

The implications of military control for civilian application of nuclear power are demonstrated by certain elements of the regime being 'transferred' to the civilian regime, 'what if?' scenarios from the military regime of control matched with their civilian counterparts:

(1) 'Should it fall into the wrong hands'... Both the fissionable materials and the scientific knowledge must not fall into the wrong hands. There exist 'wrong hands' in part because the materials and the knowledge are the source of almighty power.

(2) 'Should it be known that it's fallen into the wrong hands'... Confidence in the controllers and confidence in MAD diminishes in proportion to the knowledge of the loss of control over bomb-making capability. Loss of control should be kept secret. Analogous civilian scenarios could be:

(1) 'Should there be an accident'... Accidents lead to a lack of public trust in the system of control, only partially diminished if it can be attributed to 'human error', as in Three Mile Island.

(2) 'Should it be known there's been an accident'... Knowledge and the potential effects of military-related nuclear 'accidents', as at Bikini in 1954 (H-bomb explosion in shifted winds) and in the Soviet Union in 1957, needed to be suppressed in order to maintain control over confidence in nuclear power for civilian use. In this particular sense of the relationship, it is not at all irrational to associate nuclear war (symbolised by the mushroom cloud) with civilian nuclear power. The regime of control, with its lack of dissemination of knowledge, its secrecy, is the same.

LEGACIES OF THE ANTI-ATOMIC AGE

In the 1950s and 1960s the policy of 'public trust management' through the regime of information control was accompanied by additional confidence-building: the nuclear utopian project, which by the late 1970s and 1980s, had become a dystopian project. In the U.S. nuclear power was put under 'civilian control' in 1946, proliferating after the government's 'atoms for peace' and 'good atom' and Disney's 'friendly atom' campaigns.[06] At present, thirty countries have in operation over 400 nuclear power plants; the US has the greatest number at over one hundred and thirty.

The predictions of 'electricity too cheap to meter' fit perfectly in the age of no-limits-to-growth. Despite the unbridled use of energy, the environment would be clean; gone are the smokestacks, the smog, the dirt and detritus of the 'carbon age' (recalling the archaeologist's materials bias). Make way for the atom and the nuclear age, where not only houses, but means of transportation (cars and airplanes) and time-keeping devices (watches and the most accurate atomic clocks) would be nuclear-powered.[07] Targeted nuclear blasts also were proposed to control the weather. One idea accepted as plausible was to build small-scale, portable and operationally decentralised nuclear power plants to be used at the site of

BIKINI FLAG. EVERYTHING IS IN THE HANDS OF GOD. THE WORDS SPOKEN IN 1946 BY JUDA, LEADER OF THE BIKINI PEOPLE, TO U.S. COMMODORE BEN WYATT WHEN HE ASKED THE ISLANDERS TO GIVE UP THEIR ISLAND FOR NUCLEAR TESTING.

a (natural) disaster, providing immediate and abundant power. As envisioned in this instance in the U.S. in the 1950s, nuclear energy didn't necessarily imply a centralised system.

Accidents and disasters tend to reveal the messiness of systems, contradicting the image of rational control and efficient functioning painted to the general public and to regulators during their ritualistic inspections. Such was the case at Three Mile Island, which seemed to vindicate the claims of an anti-nuclear movement in existence since the early Seventies, with one of the first mass demonstrations occurring in France in 1971. The charge made by the protesters, pointed out by Nelkin and Pollak in their book THE ATOM BESIEGED (1981), was that "'the nucleocrats', the 'technocratic power elite', represented the nucleus of a future technology-based fascism". Certainly the radical science and appropriate technology movements, even the 'democratic technology' movements of today are outgrowths of the image and reality of a strong, centralised system of control, synonymous with the nuclear age.

Many will remember the great anti-nuclear movements in the age of apocalypse, captured in the real accidents (Three Mile Island and later Chernobyl), reports of near accidents, and 'nuclear disaster' films' as THE DAY AFTER, or in 'nuclear accident' films. The difference between disaster and accident is one of magnitude and, conventionally, the potential for change is greater after a disaster. That is, the disaster occasions discussion of system replacement, and the accident only a system redesign.

A nuclear accident film was broadcast in Germany in the late 1980s. A nuclear waste truck crashes near a small village and the Bundeswehr is called in to seal off the area. 'Contaminated' villagers attempt to flee, whereupon they're machine-gunned by the Bundeswehr. The point is that nuclear waste management, like the military-based system of civilian nuclear power, implies control not only of knowledge and materials but of the people in the area — as the inhabitants of the Pacific islands have found out. To control people in the nuclear age, secrecy is important. Nuclear shipments are not 'announced'. Greenpeace, an organisation which grew out of the founders' protests against French nuclear testing in the Pacific in the early 1970s (and was galvanised after French secret agents sank the Rainbow Warrior in 1985) and others do detective work. Owing to massive protest, 'controlling' people now costs more than the shipments themselves — better quickly bury the stuff in permanent waste dumps.

I'll conclude with the legacy of the Pacific, the white island of Bikini, captured by the Americans in their 1942-1945 Pacific campaigns, which included the bloody battles of Okinawa and Iwo Jima, when the Japanese flew their kamikaze missions. The Bikini islanders, relocated in 1946, 1947, 1948, 1969 and 1978, showered with H-bomb debris in 1954, ill and living in the shadow of the death of their friends, relatives and babies, have received payments of $150 million from the U.S. government (unlike their French counterparts) and have come to lead a different lifestyle, which the native representatives describe as one based on cash and greed. Their indigenous lifestyle is long gone, replaced by a culture of dependency. Over a half century later, they've returned home again, as they did in 1969 only to be removed in 1978 after further scientific tests concluded that the white contaminated island was uninhabitable. But what is home now? The question can be posed in terms of the environmental damage, and the islanders' physical and psychological damage, but also in terms of what the islanders are making of their home in the wake of the testing and the atomic legacies. The islanders have opted to make Bikini into an elite tourist resort, with plenty of opportunity for scuba divers to explore the shipwrecks of the nuclear age. One islander, a leading representative, said he'd like to build casinos and turn it into Las Vegas. Since atomic tourism was born in Vegas, with sunglassed spectators watching the blasts at Trinity from hotel roofs, another ironic circle may soon be closed on the nuclear age.

(01)
Hall, J., *REAL LIVES, HALF LIVES: TALES FROM THE ATOMIC WASTELAND*, Penguin, Harmondsworth, 1996, chapter six.

(02)
From the film documentary *ATOMIC CAFE*, 1982. A segment from American radio broadcasts in March 1946, when Bikini island was evacuated, and July, when the first nuclear test in the Pacific was carried out. Abel, dropped from an airplane, was a bomb equivalent to 23,000 tons of TNT, compared to 18,600 (Trinity); 16,000 (Hiroshima); and 22,000 (Nagasaki).

(03)
A couple of years ago the U.S. Department of Energy published a document entitled *CLOSING THE CIRCLE* on nuclear power. The search for permanent disposal sites — the world's first — to house all of America's civilian and military nuclear waste, and perhaps some portion of the world's, is supposedly coming to an end. If the Department of Energy has its way, the low-level waste will land in a deep salt dome near Carlsbad, New Mexico, while the high-level will rest in the Yucca Mountain site in Nevada. Both sites are quite close to where the bomb was made. Given the history of far-flung application and experimentation, perhaps it's ironic that such a decision could be a testimony to the environmentalist view that man-made waste should be disposed of at the original site of production.

(04)
See Lebow, R., *WE ALL LOST THE COLD WAR*, Princeton University Press, 1994.

(05)
Radder, H., "Experiment, Technology and the Intrinsic Connection between Knowledge and Power", *SOCIAL STUDIES OF SCIENCE*, 16, 1986, pp. 663-683.

(06)
See Weart, S., *NUCLEAR FEAR: A HISTORY OF IMAGES*, Harvard University Press, Cambridge, Mass., 1988.

(07)
See Corn, J. (ed.), *IMAGINING TOMORROW: HISTORY, TECHNOLOGY AND THE AMERICAN FUTURE*, MIT Press, Cambridge, Mass., 1986.

06

TECHNOLOGICAL NATIONALISM & RELEVANT PAST FUTURES FOR THE COLONISATION OF MARS

I can't pay no doctor's bills, but whitey's on the moon...
No hot water, no toilets, no lights [...] and now whitey's on the moon...
I think I'll send these doctor's bills, air mail special,
to whitey on the moon.
Gil Scott-Heron, *Whitey on the Moon,* 1970

In 1909 Orville Wright pulled an aeroplane out of a hanger in rural France to be flown at an air show there. It was a secret flying machine, which he'd transported and hidden there a few years before in order to prove a point. In the face of arrogance and belittling expressed by certain French aviators towards the achievements in North Carolina, he wished to prove the superiority of the flying machine he and his brother, the bicycle-makers, had developed. At the show he demonstrated that the Wright Brothers had achieved much higher standards of control and manoeuvrability than their counterparts in France, Britain, Italy and Germany. Only then did the Wright Brothers finally receive their due credit, respect within their profession and a 'just reward' in the form of a cash prize. (Patent law and 'intellectual property' were conceived with this notion of the 'just reward' in mind.) In Europe, however, the event was placed in a nationalist context; it was not the Wright Brothers, but 'the Americans' who were ahead. No longer were the newspapers and the aviation societies the only sponsors of flight, with their awards of achievement and financial incentives for the 'dashing and daring' flyers. The continental governments began putting up money for research and development into higher standards of performance, clearly realising the potential military value of controllable airplanes.

Something similar occurred on October 4th 1957, with the launch of Sputnik 1. Adding to the overall 'shock' (that led to school curriculum overhauls and subterranean building booms) was the launch one month later of Sputnik 2 with Layka the dog aboard. As with the Wright Brothers, the foreign level of technical superiority shocked. Sputnik 1 weighed eighty four kilos, Sputnik 2, five hundred and eight kilos. The 'Americans' were working on the launch of a paltry ten kilo satellite. In the late 1950s in the U.S., the question arose how the race was lost; it was attributed to President Eisenhower's separation of civil and military rocketry.

As with atomic technology, civilian authority was meant to provide reassurance of the peaceful purposes of space technology. The lost race prompted governmental support and the well-funded NASA, using Wernher von Braun's (military) Jupiter rockets, was set up in 1958. The United Nations immediately set up a committee for the 'peaceful uses of outer space', which later would lead to an international treaty.

The new races, with oxymoronic 'military peace' undertones, were on. The Soviets placed a man in orbit on April 12th 1961 and returned him safely; one month later an American was in suborbit for fifteen minutes. As with flight earlier in the century, races revolved around newer standards – of distance and duration. These were races not between project sites (as Los Alamos or Peenemunde in the 1940s) or between particular scientists per se, but between nations or blocs controlled by nations. Like the flyers preceding them, early astronauts and cosmonauts would individually become heroes. The Soviets sent the first rockets out of the earth's atmosphere, and then attempted interplanetary missions to Venus in 1961 and Mars in 1962. All were considered failures. The Soviets and Americans were racing to Mars at the same time in 1964. The Soviets wished to land a vehicle, but Mars 2 and Mars 3 crash landed in the massive dust storms. Mariner 4, with its technically superior telemetry and operational manoeuvrability, flew by and transmitted pictures back.

Why were they racing to Mars? They were in search of intelligent life, an advanced civilisation. Competively, they were looking for some confirmation of the 'grooves' and 'canals' observed by the Italian and American astronomers from the late nineteenth century into the 1950s, when friendly and unfriendly aliens hit the science-fiction books and the radio news. The pictures that Mariner 4 and subsequent Mariner missions sent back were not what the imagination had hoped for. Mars was found to be a 'cosmic fossil'.

Having failed to win the first Mars races, by taking pictures, placing a probe in Martian orbit, landing a craft and keeping radio contact with it, the Soviet Union would concentrate on living in space, in space stations. Both nations would attempt and succeed in landing vehicles on the Moon. The Moon Walk was next logical step, the next phase in the race.

The following memorandum, prepared for President Kennedy in 1961, suggests that a nation expresses itself through a technological project:**(01)**

It is man, not merely machines, in space that captures the imagination of the world. All large-scale projects require the mobilisation of resources on a national scale. They require the development and successful application of the most advanced technologies. Dramatic achievements in space therefore symbolise the technological power and organizing capacity of a nation. It is for reasons as these that major achievements in space contribute to national prestige.

According to this memorandum, a prestige project symbolises the capacity of a nation. It embodies a national goal and relies on the successful application of the most advanced technology to achieve it. The national goal behind the Moon race was winning the race itself, placing the American flag on the surface of the Moon (not unlike on the North Pole or Antarctica), and collecting material proof of presence — Moon rock. According to the UN Treaty of Outer Space (1969), "outer space, including the Moon and other celestial bodies, is not subject to national appropriation by claim of sovereignty, by means of use or occupation, or by any other means." No ownership claims were made.

Once this all had been achieved, once international imagination had been captured and prestige won, the American space program fell off accordingly. But for the fact that the astronauts brought and drove electric battery-operated cars (lunar rovers) on the Moon and

ONE WORLD TIMES, ALEXANDERPLATZ, BERLIN.

played golf, subsequent Moon walks hardly would be remembered. In fact, twelve men walked on the Moon, but the Apollo program, including Skylab, was halted in 1977 before its planned eighteenth and nineteenth missions were carried out.

Apollo had a clarity of purpose — national prestige to be accrued from winning the advanced technology race between nations. Its 'mission statement' had a certain clarity — this may be summed up in three words: man, Moon, decade. Getting there was enough. When the goal was attained in 1969, NASA scientists drew up a similarly succinct mission, again according to the 'Kennedy model': man, Mars, decade. The goal was not just to get there, but also to live there. Space colonisation has been described as a complicated science-fiction undertaking, hardly justified in the mid-1970s, now that the Moon race was won and Mariner had mapped Mars, Venus and Mercury. But the question of Martian life remained. The excitement for Martian life tapered considerable with the data collected by Viking 1, when it landed on Mars and collected surface samples in 1976. "Deader than Elvis," read the article on Mars in the WASHINGTON POST....

At that stage the Kennedy Model gave way to the Sagan model, as Robert Zubrin explains in THE CASE FOR MARS (1996). The model rested on international cooperation in the age of the Détente. The U.S. could no longer go it alone financially, as it was thought 'unconscionable' to invest billions in any scheme to colonise Mars, complete with space tugs, nuclear engines and space stations in permanent geosynchronous orbit. The 'spin-off' argument, while still referred to, didn't justify vast expenditure. (NASA still publishes a magazine listing contributions to research in "education, transportation, pollution control, rain forest protection and health care". Recently, expensive, on-demand spy-satellite images could also be listed.) Social problems on Earth would not be addressed by way of travelling to the Moon and Mars first, it was argued. The waste and extravagance were brought home by NASA's abandonment of million-dollar rovers on the lunar surface. The only Earthly justification for such missions would be Russo-American cooperation – a peace dividend at home.

The American space program still pursues the late 1960s dream of living and working in space, peacefully with the Russians, which began with the Détente and the linkage between Apollo and Soyuz in 1975.**(03)** The Space Shuttle, originally conceived as a craft to 'shuttle' between the earth, and communities on Skylab, the Moon and Mars, finally docked with a space station – Mir – in 1995. The international space station currently being lifted into the heavens is a culmination of that early Sagan model.

RELEVANT PAST FUTURES: SEVENTIES & NINETEES CASES FOR SPACE COLONISATION

A comparison of arguments for travelling and living in outer space from the mid-Seventies and the mid-Nineties shows the extent to which not just nationalism or one-earth internationalism but compelling futuristic imagery currently drives the quest for space colonisation programmes. To gain currency, the programmes are made to fit in the current cultural context; down-to-earth environmentalism in the 1970s: and end-of-history liberalism in the 1990s.

To many in the 1970s the problems on Earth were thought to be more pressing than any programme for outer space colonisation. Neo-Malthusian fears of overpopulation, increasing poverty and declining natural resources, sustained by scientific studies as *LIMITS TO GROWTH* and tracts as *BLUEPRINT FOR SURVIVAL*, prompted a popular and technoscientific movement towards the creation of a technological landscape run on 'appropriate Earth technology'. The 'futuristic space technology' movement were quick to borrow from the new ecological spirit. Either could provide solutions to the problems on Earth. While never achieved in macrocosm, which is typical, the futures continue to live with us either in microcosm, or in the particular technologies derived from the application of imagination in the period of the 1970s.

The symbolic technologies of the Earth Future movement were the garden, the windmill and the bicycle, and of the Space Future movement, the greenhouse (with 'perpetual sunlight') and the people mover. Solar power was taken up by both movements, though the former preferred a decentralised, the latter a centralised system. The Earth Future would be 'open' or outdoor, the Space Future 'enclosed' or indoor; the former suitable for living in and with nature, the latter for living in artificial environments, or 'new natures'. Though largely based on different technology, both were 'ecological' in the sense that the technological communities were to be 'self-sufficient', an outgrowth of the oil crisis.

Justified in neo-environmental terms, the envisioned ecospace colony would use natural resources available on the outer space planet, scientifically modifying planetary soil for use in the perpetually artificially-lit greenhouses in the 'agricultural sectors'. Large mirrors would direct sunlight to create naturalesque 'days' and 'nights' in the 'residential sectors'. Satellites in geosynchronous orbit would be solar-powered and the 'clean', 'renewable' energy beamed back to Earth. The neo-Malthusian appeal was that space colonisation (combined with 'appropriate', planned parenthood) could eventually alleviate overcrowding on Earth. Living in space was thus made ecologically defensible in the cultural context of the 1970s.**(02)**

Addressing earthly concerns, in the colonies there'd be no cars – the recognised source of pollution and development ('sprawl'). People would be moved by horizontal escalators (people movers) and electric monorail trains (as at today's airports). The technology and architecture of contemporary shopping malls (the Peachtree Centre in Atlanta and Portman's Bonaventure in LA) similarly resemble Seventies' depictions of space communities as does the underground living complex in Crystal City next to the Pentagon (not to mention the Washington, DC metro). Overlapping with the nuclear age, in the space age there were any number of technological and architectural dry runs, combined with lifestyle image-making, on Earth. (See also **SEVENTIES NOW!**)

The recent Biosphere projects reveal the contemporary progression of the ecologically self-sufficient Space Future vision, even if it seems to be more associated with the Appropriate or Whole Earth movement these days. In space, at least for the U.S., only the Shuttle remains as a reminder of the imaginative Seventies 'past future' of space colonisation. The other side of the space program – space commercialisation – has never captured the imagination unless tied to a grandiose vision like instant world-wide universal mobile telephony, recently reinvigorated by Motorola's Iridium and its competitors, ICO and INMARSAT.

The Nineties case for space colonisation (and terraforming) is made most forcefully by Robert Zubrin, and has been popularised in neo-liberal terms. When compared to the Seventies case for Mars, the Nineties case points out the context dependency of arguments made to achieve essentially the same technological mission – man colonising planet. The Nineties case also shows how earlier models (relevant pasts) are employed as 'guides' to make current futuristic cases more compelling. To make a case for a futuristic technological project, the promoter often must find 'usable pasts', or indeed, 'usable past futures'.

For Nineties would-be Martians, the usable hero isn't Armstrong or Glenn, or Big Science – Los Alamos or NASA – but Amundsen (North Pole) and Shackleton (South Pole) and their small-scale private undertakings complete with image-making media campaigns aimed at the emerging popular yellow press of the turn-of-the-century. It's been called not the Kennedy or the Sagan but the Newt Gingrich model, Zubrin points out in THE CASE FOR MARS. As ever men will yearn for fame, gain and the satisfaction of curiosity. The gain this time around would be a cash prize, similar to those earned by flyers and the pole explorers after the turn-of-century. Zubrin called for the creation of a $10 billion prize for the successful pioneer of the new Martian frontier.

If the drive behind a mission to Mars in the 1960s was 'science fiction science' and the prospect of extraterrestrial life (see also Pioneer 10's and 11's plaques of the early Seventies and the Voyager's gold-plated copper disc of 1978), if in the Seventies it was reformulated into the creation of the ultimate test-bed for ecological self-sufficiency in the outer space technological landscape, then the popularised Nineties case sounds tawdry by comparison: 'Mars for Commerce', mass television 'entertainment', 'adventure travel', an 'exploration saga', reminiscent of the Vikings, and the pole explorers. 'Mars for fun and profit' headlined an article in the *NEW YORK TIMES MAGAZINE* (26 May 1996). Zubrin's piece in *TECHNOLOGY REVIEW* (November/December, 1996), 'Mars on a shoestring', refers to a private venture much leaner than NASA's 30-year, $450-billion project, put forward in the 1980s. The private frontiersmen are willing to take Amundsenian and Shakletonian risks for fame and gain.

Note how the installation of a research station on Antarctica was an earlier earthly testbed for outer space colonisation, in this passage about the future of Martian colonisation from the *NEW YORK TIMES MAGAZINE*:

> *The Antarctic's recent history, in fact, may be the best guide to Mars's future. Although it failed to become a hot spot for colonisation after being charted by the polar explorers, Antarctica has nonetheless been a boon to humanity. Government scientists, taking advantage of the risky trailblazing done by private explorers, have set up research stations that have yielded vital data for today's environmental debates. And another group of people has followed the explorers. Arriving in cruise ships during the Antarctic summer, clad in identical bright red parkas, these visitors clamber on the white cliffs and take pictures of one another posing with penguins.*

The 'identical bright red parkas' harken to depictions of pioneer space colonists. Once settled in, according to the science fiction science literature, they normally don light-weight clothing, William Morris *NEWS FROM NOWHERE* style, breathing an air of sexual freedom not found on Earth. Club Med on Mars.

The experiment with the first tourist in space, or non-astronaut/cosmonaut, ended in tragedy in January 1986. It was later argued before the Rogers Commission investigating the Challenger Space Shuttle disaster that putting a teacher in space was scandalous. Since Nixon, NASA had backed down from the 'prestige project', rhetorically turning the shuttle into 'routine' travel. Indeed, the tragic episode would seem to call into question the routineness with which tourists or actors could travel to Mars, with their (adult) exploits beamed down to earth for Soap Opera viewers.

Another segment of interest refers to well-worn ideas about the outcomes of frontierism, and the impending end of history:

Having a planet of refuge could be of enormous social use. It could reinvigorate civilisation on Earth with its new ideas, just as America reinvigorated Europe. In human history, free dynamic societies have existed only during the four centuries of frontier expansion in the West. Without a new frontier, our civilisation is doomed to stagnation.

Like the NASA programs following Apollo, Martian travel is a program in search of contemporary justification and an historical analogy, a past to drive its future. The author has returned to the age of speed and nationalism for a promotional cause for his case.

The subsequent discovery of primitive life on early Mars will bring craft back to the red planet. As we plumb the past for inspirational models for the present, we'll surely read about nineteenth-century astronomers, H.G. Wells and Orson Wells and Ray Bradbury, who wrote in the *MARTIAN CHRONICLES*, and also said when Viking 1 landed on Mars, "We are all Martians now". The future probe will have to dig deep into the Martian surface to determine if the rocks that landed in Antarctica many years ago carried life from Mars, or had been 'contaminated' by life from earth. It is also sure to revive the imagination for further deep space probes, as we recall how the Pioneers and the Vikings are plunging deeper and deeper into outer space with messages in bottles (or on CDs) from Earth onboard. Contrary to the individualist, swashbuckling space explorer model, we also may recall the age of Esperanto, when linguists and futurologists, inspired by space internationalism, persevered to transfer the one universe ideal down to Earth.

(01)
Jensen, C., *CONTEST FOR THE HEAVENS*, Harvill, 1994, p. 73.

(02)
See NASA's *SPACE COLONISATION: A DESIGN STUDY*, 1976 and *NATIONAL GEOGRAPHIC*, July, 1976. See also Brand, S. (ed.), *SPACE COLONIES*, Waller Press, San Francisco, 1977.

(03)
See McCurdy, H., *SPACE AND THE AMERICAN IMAGINATION*, Smithsonian Institution Press, Washington, DC, 1997.

07
FLIRTING WITH THE GHOST IN THE MACHINE

HUMANKIND has long striven to remake itself in its own likeness. Over the course of history humans have employed three distinct types of force to achieve varying degrees of human likeness — bodily, (electro)- mechanical and electronic. The puppeteer and the ventriloquist, for instance, exerted the bodily, making the inanimate, human-like objects appear to come to life, investing them with motion and speech. In their performances they often were able to provide their creations with feeling and emotion, for the creatures were direct extensions of their operators. Though they attempted to make their creatures stand alone through affecting a distance by sleight of hand, both the puppeteer and the ventriloquist were connected to their 'wooden' human likeness because of their direct human control. That connection provided the audience with a sense of security, at least upon reflection.

The clock-work automata of the eighteenth and nineteenth centuries did stand alone, however. Once set in motion they were no longer connected to their operator, they were 'unleashed'. A number of insecurities grew up around the automata, especially when they were equated with machines that did not 'act' as they were supposed to. Through mechanical failure they could go 'haywire' (a 1905 coinage), like the feeding machine in Charlie Chaplin's *MODERN TIMES*. Once 'unleashed', the wound-up automata — the machine in human (or animal) likeness — could literally 'spin out of control'. This is one of the powerful origins of the notion of 'autonomous technology', and the fear, or insecurity, surrounding technology 'out of control'.

GHOST IN THE MACHINE

Frankenstein's Monster is to science what automata and robots are to technology. Unlike the Monster, however, the automata could display 'humanness' only superficially, through a painted-on smile or through sentimental human-programmed responses. The machine had an uncontrollable, autonomous 'spirit' only in our imagination. Not too many automata were destroyed for fear they had 'minds of their own', as they were in a Stanford experiment, when participants had to choose between smashing a cuddly animal-like robot or an insect-like robot. Electronic force added a new chapter to the potential autonomy, or 'self-government', of the machine, and the perceived loss of direct human control over it — where control is defined here as masterful (human) operation coupled with predictable (machine) response.

The dawn of electronic machine intelligence arrived during World War II, if we define the puzzle-solving involved in code-breaking, narrowly, as a form of intelligence. The 'intelligent machine' made its public appearance a few years later. When in 1952 the UNIVAC predicted a landslide victory for Eisenhower in the televised coverage of the presidential elections, a number of elements of the autonomous intelligent machine came together: the computer was granted the capacity of intelligence when it was 'asked' to predict the results (not by the great swami, the magician, the conjurer or the amusement park impresario, but by 'serious' men in suits and technicians in clean lab coats.) The UNIVAC produced an unpredictable response, for the experts disagreed with outcome of the calculations: a landslide victory for Eisenhower over Adlai Stevenson. The machine acquired an expert soothsaying ability, providing a sense of wonder more in common with the automata of amusement parks and fairs. Input was coded in, lights flashed, and the UNIVAC prophesied the future. Because the 'experts disagreed', because they were 'outsmarted', the machine seemed to have a mind of its own; the results seemed to be of its own making.

One important outcome of the UNIVAC television experiment was that the 'intelligent machine' could be trusted to provide information to form the basis for certain decisions. The flipside was that the experts — mathematicians and statisticians — also learned that one disagreed with the intelligent machine at one's peril. They could appear fools beside the machine. This has been of relevance for the development of 'decision support systems', and, more radically, 'expert systems', whereby the machine is granted a status of 'the thinker' with a capacity to generate truth. This capacity may be deemed superior to human non-experts', even to those who feed the machine (with data). With the introduction of the expert system, even certain experts may become less expert in the workplace; their expertise may be challenged and undermined by the codification of expertise built in the software of the machine.

'Resistance' to expert systems is often explained in superficial historical terms as the fear of the unknown; variations on the theme of machine resistance are couched in the opposition between man and machine (e.g. Kasparov vs. Deep Blue). It may be more telling to state that systems 'granted' a potentially superior capacity to generate truth are resisted, especially if the human is required to conform to the decision-making regime based on the 'mechanical brain' — or artificially intelligent machine — over his own decision-supporting expertise. More specifically, the systems may be resisted because the operator is faced with a dilemma upon seeing the output, whether to agree with the machine and to act accordingly, when intuition and expertise may suggest otherwise. The obvious way out of the dilemma is troubling as well. If you agree with the machine and you're wrong, you can blame your decision on the machine, with some measure of impunity — which immediately brings you to the historically dubious rationale that you were just following orders, or 'instructions'.

Machines cannot be prosecuted. The potentially liable manufacturer and/or programmer will tend to disagree with the operator's accusation, however, and argue that it was improperly programmed, operated or understood. The opposition, in other words, is between humans interpreting human-machine interaction, and not between humans and machines. A future match could be between human with expert chess system, and Deep Blue. Then the analysis will be of when and how often the human player disagreed with the expert system — listening to his voice — and when and why one was seduced by the ghost in the machine.

At 'Doors on Tour', sponsored by the Netherlands Design Institute in 1997, Willem

Velthoven (of Mediamatic) talked about 'pure interactivity' in contrast to 'user friendliness' in the context of website design. Pure interactivity would require the machine to talk back and challenge users to think and react. The machine wouldn't just smile back 'stupidly', like the Apple Mac always has, but show different sides, and act less 'friendly', more 'unpredictably' — with more 'personality' and wit. This conjured up images of the Turing Test and (artificially) intelligent programs (on display in Boston's Computer Museum) as ELIZA (1966) which employed the Rogerian method from psychoanalysis.(01)

— Men are all alike.
— IN WHAT WAY
— They're always bugging me about something or other.
— CAN YOU THINK OF A SPECIFIC EXAMPLE
— Well, my boyfriend made me come here.
— YOUR BOYFRIEND MADE YOU COME HERE
— He says I'm depressed much of the time.
— I'M SORRY TO HEAR YOU ARE DEPRESSED

At Doors on Tour the call was for the return of the ghost in the machine, the return of the ability of the 'clever machine' 'to outwit' the human. We do often flirt with conjuring up a ghost in the machine, mainly so we supposedly won't have to work as much or get as dirty, as in the case of the one major early experiment with science-fiction style robotization in an American car factory — General Motor's in Lordstown, Ohio in 1971. It went awfully wrong: "boring jobs, high absenteeism, workers turning to drugs and alcohol to escape meaningless jobs, even sabotage to get back at the machines that seemed to be controlling their lives".(02)

One may define the call for the return of the ghost in the machine, for 'pure interactivity', as a 'paradigm challenge', as a challenge to the guiding principle of user friendliness — the Apple construct, with deeper roots.

THE NEW GHOST IN THE MACHINE IS THE ENVISIONED USER

While the movement goes back to the early 1970s, the symbolic arrival of user friendliness — and mass computing for all the people in the U.S., and not just the mathematicians, statisticians, experts and male nerds — was probably the Apple television commercial broadcast, only once, during the Super Bowl in 1984. In the commercial, Big Brother of 1984, representing bureaucratic, regimented Big Blue (IBM), was staring down at mass-production workers from a giant television screen. A woman wrests herself from the work regime and hurls a hammer, smashing the screen, and all that went with it.

Freedom from 'enslaving technology' through new 'convivial technology' — smash the screen which controls you — you, the producer and you, the user. But who is this user? Blue-collar workers still mass produce the machine, skilled technicians some of its components, on the West Coast, Asia and beyond. Is he or she controlled by the machine, (cognitively and physically) forced to conform to it, in a Foucauldian sense? Is s/he included in the machine or excluded from it? Where do inclusion and exclusion occur and how do they work? The paradox of the Apple commercial was that the Apple Mac was made for (and perhaps itself made) a new mass of workers. A ghost also was inscribed into the machine, but she, the envisioned user, wouldn't be operating it.

In the mid-1980s the *WHOLE EARTH REVIEW* (a West Coast magazine published since the late 1960s) printed an advert for The Well, the archetypal on-line community described by Howard Reingold in *THE VIRTUAL COMMUNITY*: "To join The Well, you don't have to be a computer person, just a person with a computer." Such a sentence was only just possible in the mid-1980s, for computer persons — either the employee-end user, hackers or the amateur computer operators, analogous to their radio counterparts in the late 1910s and 1920s — were no longer the only ones with computers.

Like the radio amateurs of old, early computer amateurs put machines together themselves by ordering parts or complete kits. The wooden Apple I was a product of a club of computer amateurs — the Homebrew Computer Club, and their enthusiasm for the organic, counterculture, community and The People. Not only was the 1984 marketing of the Apple Mac ideological, but so was its design: the desktop graphical interface was a design for the people. The Apple Macs supposedly delivered 'empowering technology' for the populous at large, just like citizen's band (CB) radio of the same period, used mainly by 'communities' of automobile drivers to chat and warn each other (in arcane language) about the location of the State (highway patrol cars with radar guns aimed at impeding the freedom of the road).**(03)** The Apple Mac was empowering in the sense that it would allow the non-computer person to compute, it would liberate the user from learning arcane programming code. It would be many years, however, before the archetypal user actually worked with its ideological interface.

The developers of the graphical interface (at XEROX Parc) had Sally, the non-technical secretary in mind, not interested in the inner workings of the machine, but only that it worked and could be operated with ease.**(04)** The icons on the screen — files, folders and the trash can — resembled the environment of the office, and were thus comprehensible and comforting. It is sometimes claimed that the interface allowed the secretary to understand how the machine worked.

The ability to run more than one application at a time ('multi-tasking') derived from the notion that the secretary always had to do more than one thing at a time. It could be called non-linear peripheral work, in contrast to linear vertical work, with the appropriate typical female and male gender styles respectively attached. In the terms used in gender and technology, the represented (or preconfigured) user was the non-technical, desked female; one 'gender script' (i.e. 'the real-life play written in the machine') would be that secretaries will use computers designed for secretaries and remain secretaries-now with computers.

(While circular, the point is that new technological designs inscribe, and the implementation of the design machines reify, (corporate) gender relations.) Most business secretaries, however, would never touch an Apple Mac, for the IBM PC arrived just in time. Big Blue was back in new business with their personal computer – computer for the person. Sally now can work what Apple inspired: Microsoft's Windows.

Apple, however, 'scripted' something else. It 'blackboxed' its operating system, so peripheral, intuitive tinkering only took place on the level of the graphical interface and not at the level of the system code. The IBM PC came with Basic and other programming languages (as FORTRAN) and AT&T's UNIX and later 'C') could be run on it as well. Applications could be written and run on the standard disk operating system. DOS always has been open, explaining why many amateur computer operators detest the closed Apple system. Apple iron-caged its core, though later the company threw open parts of its OS to allow 'third parties' to develop applications, uninvited.

Though it may seem so, the Apple Mac was not 'designed for all'; amateur computer operators – men, mostly – were excluded 'by design for the masses', where the 'mass user' means lowest common denominator user – a living, working stereotype of the non-technical female. The secretarial gender script of the Apple 'disempowered' the typical male programmer-user. Conventionally, this happens as a technological design begins to mature for the mass market; manufacturers tend to protect the inner workings from playful amateurs, hackers and the uninitiated 'inner users', whether adults, adolescents or children. Neither the Apple casing nor the software has invited tinkering, intuitive thinking, at the level of the operating system.

COMPUTER UNDERSTANDING & USER EMPOWERMENT

Moving away from gender, I would like to pursue the notion of 'empowering technology' a little further. Apple assumed that one needn't understand the inner workings of the machine and the inner codings of the software (through closed architecture and OS) to feel empowered by the technology. 'Expert operation' (not expert understanding) implies empowerment, and it can be achieved by 'dumbing down' the control interface, by making it accessible to the non-technical expert. Indeed, this operative empowerment would be in keeping with the everyday skill of the videogame player, and the operator of simulators, or the real thing (say, of missiles), the interfaces of which increasingly derive from the most successfully operated simulators. Once debugged (an 'expert user' feedback process), the program and the machine can be mastered by the operator, whose mastery grants him or her a sense of power not over but with the machine.

Generally speaking, this Nineties viewpoint concerning technological empowerment contradicts the (counterculturally adopted) Sixties philosophical idea of human alienation by technology through incomprehension of the machine (which may enslave, make you conform, violate your will), above all threatening the human condition, and not empowering humans in some cognitively or socially beneficial sense.

As we enter an age of smart technology – a time when technologies learn user behaviours and preferences and act more and more in accordance with previous patterns of use – alienation more likely will derive from technology making you conform not to the designer's envisioned user but to your old self.

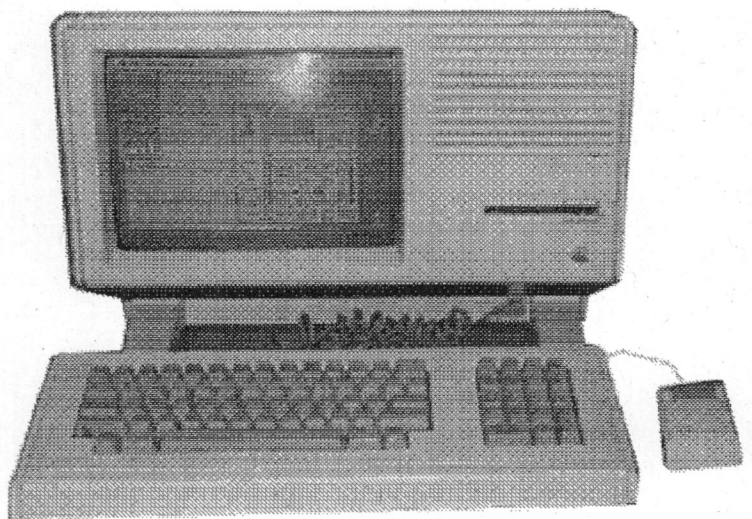

THE FIRST FRIENDLY COMPUTER. APPLE LISA, 1982-83. $10,000. SOURCE: COMPUTER MUSEUM OF AMERICA.

(01)
Weizenbaum, J., *COMPUTER POWER AND HUMAN REASON*, W.H. Freeman, San Francisco, 1976.

(02)
Lubar, S., *INFO CULTURE,* Houghton Mifflin, Boston, 1993, p. 331.

(03)
For a history of The Well, see also the May, 1997 edition of *WIRED*. For the 'virtual community' which grew around the telegraph, see Standage, T., *THE VICTORIAN INTERNET*, Weidenfeld & Nicolson, London, 1998.

(04)
Van Diemen, D., "De Facelift van Sally, Doorbraak van een Nieuwe Computer Cultuur", *LOVER*, 3, Amsterdam, 1997, pp. 18-31. See also T. & A. Hornath, "The Social Construction of The Personal Computer User", *JOURNAL OF COMMUNICATION*, 45, 3, Summer 1995.

08
SEVENTIES NOW! MODERN & POSTMODERN TECHNOLOGICAL DESIGN

VIEWED from the history of technological culture, the spirit of Modernism began underground, swept the surfaces of the western world and ended up in the heavens, in outer space and, finally, in cyberspace. It climbed the Apollonian vertical axis of progress, and came crashing down to earth.

In 1970 Herbert Marcuse defined utopia as the "determinate sociohistorical negation of what exists."**(01)** Negation, however, does not mean denial but recognition — that we don't want what exists. The negation of what exists by planning to change it by total (or totalising) design for all — this is the 'Modernist spirit'.

What would modernists and postmodernists want, and what would they want to negate? To answer this I will focus on the turning points from the pre-modern to the modern beginning in the 1860s, and from the modern to the post-modern in the 1970s. For some (see Stephen Toulmin's *COSMOPOLIS*) modernism began with the Enlightenment, but here I'm thinking about technological design, not science.

The 'symbolic beginning' of the age of modernism can be sited in the reconstruction of the Paris sewers in the mid nineteenth-century, as described by Victor Hugo.**(02)** The project turned the small subterranean 'maze' or 'labyrinth' (Jean Valjean), into a great 'neat, cold, straight, correct' sewage system. It 'regularised' (Haussmann), even 'tamed' the unruly underground, eradicating the mythic in favour of the rational. Hell even became a slice of heaven, when the white-clad tour guides ushered the bourgeoisie through the new clean system during the International Exposition of 1867. Instead of wild nature, the well-heeled now toured a planned technological totality, and liked it.

From its inception, Modernism can thus be viewed as the giant 'clean-up', the great 'public health and welfare' ordering project, begun underground first in Paris, then in London, New York and elsewhere. Initially human and animal detritus was swept off the streets and channelled into neatly buried tubes.**(03)** After the waste, its producers followed — people, horses and carriages were then ordered underground, in two senses of the word. Like the sewers, the subways negated the existent surface chaos through orderly subterranean channelling. The early subway tubes in Paris, London, New York, Boston and Budapest were relatively shallow lines following the city streets they were meant to clear.**(04)** They were not straight, but were made to seem that way.

Modernism thus was both an actual and representational, or graphic design, project. As the subways grew in depth and scope, the subway systems were increasingly depicted on official maps as not following above-ground streets, but a modernist subterranean logic: straight lines from underground station to underground station. The principle behind this representation was flow, or streamline. The term itself thus reflects the giant clean-up project. The streams of sewage and the lines of people — all must flow. Representational modernist technological design was born underground.

AS BELOW, SO ABOVE

With street people and their street mess cleansed and modernised underground, the orderly and formalist beautification of the above ground city could begin in earnest. Like waste, anarchic chaos – the bombers, assassins and their sympathisers – was cleared through tearing down the milieu in which it would cluster, fester. Their cells, where the poor also lived, gave way in project after project to avenues, trees and electric light – the great (mental) cleansing and vivification agents. Thus it became difficult for the chaotic ones to 'go underground', since it had already been modernised. The modernist project, it seems, had original political purpose. In Spain, Franco also recognised the value of wide streets and even wide corridors in the universities, the seedbed of chaos. In a modernist planning scheme, there's no place to hide.

Modernism, then, is the gardener's vision. Planning and conscious design will achieve perfection by weeding and end with artificial order. Bauman described the process of the 1930s and early 1940s:

> Stalin's and Hitler's victims were... often killed in a dull mechanical fashion with no human emotions – hatred included – to enliven it. They were killed because they did not fit, for one reason or another, the scheme of a perfect society. Their killing was not the work of destruction, but of creation. They were eliminated, so that an objectively better human world – more efficient, more moral, more beautiful – could be established. A Communist world. Or a racially pure, Aryan world.
> In both cases, a harmonious world, conflict-free, docile in the hands of their rulers, orderly, controlled. People tainted with ineradicable blight of their past or origin could not be fitted into such unblemished, healthy and shining world. Like weeds, their nature could not be changed.**(05)**

MODERN ROBOTIC WORK

With communism, socialism and social democracy came the distinction between 'onerous toil' and 'pleasurable work'. The seminally symbolic (and real) site of onerous toil was the mine. In previous ages, mines were worked by slaves. As that practice became unacceptable free men would work mines, though they were still, in one sense, viewed as slaves through the widespread call for machines to replace man-slaves (miners), employed to do 'onerous toil'.**(06)** Machines should liberate men and allow them to undertake pleasurable work and leisure activities.**(07)** The celebration of the machine and the machine aesthetic, the admiration for speed and efficiency, was born of the desire to ban that particular form of slavery.

As fewer men took to the mines, more men (and women) began not working but 'toiling' in factories. The high-low axis, with its connotations of master and slave, still held. In Fritz Lang's *METROPOLIS* (1926), the man-slave machine tenders operate in the bowels of the technological complex, while the fat cat watches his television screen and operates his controls from above. Mechanisation had taken command, to cite the title of Siegfried Gideon's influential book (1947), producing a different sort of onerous work. The machine did not conform to man; man conformed to the machine, as Charlie Chaplin showed in *MODERN TIMES*. The recognition that conforming to the machine created onerous toil had a particular result. This was not to humanise the factory, but to dehumanise it further. The quest for full automation – the quest for the fully-automated factory, realised in the early 1970s, is hypermodernism achieved.

From the perspective of technological culture, hypermodernist design has at its core a particular principle: the single command-and-control centre. This explains why Jeremy Bentham's *PANOPTICON* has become the metaphor of modernism, described by Foucault. The clock of the early nineteenth-century utopian factory towns controls because it's the central point of reference, but it itself cannot see. But contemporary operations centres of large technological complexes, like a fully-automated factory, a nuclear power plant or NASA, contain central command posts that can be seen and that see, a concept borrowed from the military (i.e., the chain of command) and from the prison (i.e., the command of the chained).

In the early Seventies the Benthamite centre was the mainframe computer on which models of whole systems were tested and eventually applied.[08] Complete knowledge of a system was sought and assumed possible to have: model it and control it by planning to perfect it. This desire for total control began with waste, moved on to people and then machines, viewed as extensions of man – machine-slaves. Slaves can revolt and machines, as it's been feared since the automaton, could become 'autonomous'. Both require control.

Larger and larger systems, as in die autogerechte Stadt ('city for cars'), soon became the focus of control projects.

AS ON EARTH, SO IN HEAVENS

One could argue that attempts to totally control not waste, people, machines or people-driven machines in the city, but 'nature' and the 'universe' ultimately undid Modernism in the 1970s. Two examples stand out. The first involves a most unpredictable system: the weather and the climate. Weather modification began with cloud-seeding or 'rain-making' the late 1940s and 1950s, and reached its height during the Vietnam War, when U.S. pilots seeded the clouds above the Ho Chi Minh trail. It's ironic that the underground networks dug by the Vietcong put paid to U.S. dreams of controlling the surface of the earth from the heavens. Thus the pre-modernesque secret labyrinth, in the style of *LES MISÉRABLES*, upset the modern. The untidy East undid the tidy West. U.S. Government-sponsored weather modification was cancelled in the mid-1970s.

The other example is outer space. Here the plans of the Soviet Union, the greatest of modernists, stand out. For modernist space enthusiasts, the achievement was not the man on the Moon or the Pioneer and Voyager missions deep into the universe but the first robot with wheels and eyes to 'walk the Moon', accomplished by the Soviets on 17 November 1970. The robot scooped up some Moon dust, returned to its ship, and via its ship to Earth, all remotely controlled from the command centre.[09] 'Progress on wheels', was how the DAILY MAIL put it. For the modernists, 'progress' means the elimination of human unpredictability. The modernist spirit negates the existence of humans.

The robot was but one preparatory mission for the grand aim of colonising the Moon and Mars, a program which reached its height in the mid-1970s. But the symbolic end of technological Modernism was when Skylab, the U.S. space station, intended to form the link (with the Shuttle) between Earth and colonies on the Moon, burnt up in the earth's atmosphere in 1979. The ultimate goal of controlling or subduing the heavens fell to earth. (When someone says he's "down to earth" I assume he's against robots and space travel. He's certainly against Modernism.)

FOUR CHEERS FOR POSTMODERNISM

The adjectives 'modern' and 'postmodern' refer to periods in time, or meta-ages. 'Modernist' and 'postmodernist' refer to styles of thinking. The principle of eliminating or negating human unpredictability through total control is the modernist style of thinking embodied in technological design.

We are accustomed to hearing and believing that postmodernism signals the end of the 'grand narratives', the totalising ideologies beyond liberal democracy. According to social and political thinkers, because there seems to be no alternative to liberal democracy (apart from Islam), we have reached 'the end of history' — it was one of the first postmodern cheers. The fall of Soviet Union marks the beginning of postmodern politics.

In architecture and design circles, it is widely held that the transition from the modern to the postmodern commenced with the demolition of the award-winning, Corbusier-style Pruitt-Igoe housing complex in St. Louis, demolished to cheers in 1972. Humans can't live in machines for living, we are not robots. Thus 'Einstuerzenden Neubeuten' — another postmodern cheer. Portman's Bonaventure marks the beginning of postmodern architecture.[10] The student union at the University of Twente marks the beginning in the Netherlands. In both the design makes the visitor lose his way; it produces disorientation. The student union was designed to force the reserved eastern Netherlanders into conversation with outsiders asking for directions.

The postmodern spirit extended well beyond architecture, however. Perhaps it's best summed up in our third postmodern cheer: Earth First! But apart from the thinking of the neo-Malthusians and the 'counterculture', why should the Earth come first, first before what? Before the Heavens and Outer Space, of course. Down to Spaceship Earth. The cancellation of the grandiose space program would mark the beginning of the postmodern Earth environmentalism.

THE CENTRE MAY NOT HOLD

In search of postmodern technological design, the internet has been the prime candidate, not because it's supposedly 'chaotic', but because it has no centre, so single command-and-control point. It has no centre by design: C3I — command, control, communication and intelligence — was how the U.S. military phrased the core of its operations. It fit the larger Cold War military policy of spatial 'dispersal', which viewed the traditional city as dangerous and endeavoured to render density obsolete.[11] ARPANET was so designed that C3I could not be destroyed in either a single or multiple atomic blasts, a distributed network that would allow every surviving subcentre of operations to communicate with each other. ARPANET decentred C3I. Broadly speaking, however, the new technological design did not have the 'effect' of changing or rethinking the military hierarchy, it wasn't supposed to.

The second potential difference between modern and postmodern design is that postmodern design is meant to reflect human unpredictability (and in the case of the Bonaventure and the student union at Twente) even force human unpredictability. The notion that one can alter culture, quick, by technological design was the fatal flaw in the ideological project called Modernism. When postmodern design isn't flexible enough to reflect, but is inflexible so as to force confusion and chaos, it falls into the Modernist trap.

Popular culture is now enamoured of the Seventies, but there were at least two Seventies — those of the leisure suit and easy tunes and those of the grander and grander systems of control. Seventies Now! is the cheer of the new generation, reared on both.

This essay is adapted from a lecture given on 'Wim Crouwel Day', on the theme of Total Design, Netherlands Design Institute, Amsterdam, June,

[01]
Quoted in Sorkin, M., "See you in Disneyland", in *VARIATIONS ON A THEME PARK: THE NEW AMERICAN CITY AND THE END OF PUBLIC SPACE*, Hill & Wang, New York, 1992.

[02]
Reid, D., *PARIS SEWERS AND SEWERMEN*, Harvard Univ. Press, Cambridge, Mass., 1991.

[03]
New York Public Library, *GARBAGE! THE HISTORY AND POLITICS OF TRASH IN NEW YORK CITY*, New York, 1994.

[04]
Bobrick, B., *LABYRINTHS OF IRON*, Newsweek Books, New York, 1981.

[05]
Bauman, Z., *MODERNITY AND THE HOLOCAUST*, Polity, Cambridge, 1989.

[06]
Illich, I., *TOOLS FOR CONVIVIALITY*, Calder & Boyars, London, 1973.

[07]
Bookchin, M., "Liberatory Technology", in *POST-SCARCITY ANARCHISM*, Ramparts Press, Berkeley, 1971.

[08]
Kwa, C., "Modelling Technologies of Control", *SCIENCE AS CULTURE*, 4, 20, 1994, pp. 363-391.

[09]
Harvey, B., *THE NEW SOVIET SPACE PROGRAMME*, Wiley, Chichester, 1996.

[10]
Jameson, F., *POSTMODERNISM, OR, THE CULTURAL LOGIC OF LATE CAPITALISM*, Verso, London, 1991.

[11]
Soojung-kim Pang, A., "Dome Days", in Spufford, F. & J. Unglow, *CULTURAL BABBAGE. TECHNOLOGY, TIME AND INVENTION*, Faber & Faber, London 1996.

09
LISTEN, CITIZEN

IF our age has supposedly jettisoned the master narratives, the modernist plans for new world order — perfect gardens and technological landscapes weeded and shorn of human unpredictability — haphazard liberal democracy is the form of governance and social intercourse that has endured the recent reshuffling. In our new times of 'post-ideological niche pastiche' no other form is considered, as is sometimes said. For some, the impression is that we are careening into the Bimillennium, with liberal rudderlessness. Indeed, for the year 2000 there seems to be no urgent social agenda, only an urgent entertainment agenda.

Yet, we are continuing the search for the 'past futures' deemed most relevant for new times — the greater or lesser utopias, or parts thereof, which people rewrite in order to recast scenarios of the future. We learn the past futures for at least two reasons. They aid us in thinking through the ideals, principles and social relations which have been and could be reflected in and designed into our technologies, bringing within our grasp the ability to 'imagine alternative technological designs' and act accordingly. Secondly, comparison is the stuff of case-building. Drawing the right parallel (or spotting the spurious analogy) is one step in proposing or opposing particular cases to be made for new technology and new forms of decision-making on technology.

In *DEMOCRACY & TECHNOLOGY* by Dick Sclove (1995) the search for relevant past futures seems to be completed.**(01)** Sclove has resurrected and pitched the ideals of strong democracy, as practised in varying degrees, in the Old Order Amish communities in New England and in Swiss villages and cantons. Other successes in the Basque country and Emilia-Romagna may be more familiar to those who have read the technology and political economy literature. The ways of Ancient Athenians, Volvo in Kalmar, Sweden, the Boimondau 'Community of Work' in France and the Amazonian Mundurucu Indians are further guides, as are experiments and institutions in Canada, Denmark and the Netherlands, including participatory design and consensus conferences. To greater or lesser extents, these are the beacons of strong democracy, the cases of citizen (and worker) participation in the design, development, organisation and/or assessment of new technology which hold up a new grand narrative. The foundation is the ideal citizen and small-scale community (which the Unabomber also desired), engaging constructively in politics (and the politics of technology), self-managing their workplaces, and self-actualising themselves.

The overall guiding model of governance, social intercourse and technology decision-making — our relevant past future — is still practised by the Amish who have not succumbed to the proverbial end of history:

> *The Amish have for centuries guided their social and technological development with unusual self-awareness, while retaining their communities' self-governing, democratic character. Their success suggests that the Amish have learned better than other societies to perceive technologies' nonfocal, structural dimensions. For instance, their reasons for favoring horses over tractors include recognition that*

horses reproduce themselves; produce manure fertiliser; compact the soil less; and help avert dependence on petroleum products, parts suppliers, and outside mechanics... [T]he Amish observe the social effects of technological innovations in surrounding non-Amish communities and then use those observations as one basis for making their own decisions. (p.57ff)

Just as the Amish together learn about and question whether each new technology is 'good for the Amish', accordingly accepting it, rejecting it or putting it on probation, so local communities in western society should first ask, is it good for democracy?**(02)** Whereas (Sclove argues) debates on technology normally focus on cost, feasibility, safety, environment and sometimes national security, in the new utopia the primary criterion for the acceptance or rejection of the technology would be whether it promotes and sustains egalitarian community and conviviality. The relevance of old or existing technologies also should be assessed, and collective communal decisions taken. 'Authoritarian technology', defined as that which helps to "establish or maintain illegitimately hierarchical social relationships", should be rejected. Communitarian/cooperative technology for home life and workplace, which "help to establish or maintain egalitarian, convivial, or legitimately hierarchical social relationships", should be furthered.

Not only at work but also at home, authoritarian mass-production regimes impede the development of conviviality. As Sclove argues: "Authoritarian work lives, individualised home lives and mass-consumer-spectator lives" reproduce one another. Mass-production firms (where the worker conforms to the regime of control) wish to sell individualised technologies (a washing machine in every home), which thereby eliminate the convivial interaction, at the laundromat. Televisions or stereos in every home, or in more than one room of the home, produce similar non-conviviality, or radical individualism." (p.66)

Apart from discussing the (varyingly cooperative) craft and cottage industries which mass production replaced in most countries, we haven't gone into convivial technology, which has a lineage in the history of ideas on technological culture. It should be recognised that convivial, or democratic technology, has been defined before, for instance by Henry Ford.

From Ford's cars to Apple's computers, "democracy," according to Sclove, "became equated with a perceived tendency toward equality of opportunity in economic consumption". If there is equal opportunity for its consumption (the culture of Fordism's Faustian bargain), then technology is democratic. Sclove rejects this self-serving definition of 'democratic' technology — technologies which further radical individualism (and are produced in mass production workplaces) are construed as non-convivial.

'Conviviality', according to Ivan Illich in *TOOLS OF CONVIVIALITY* (1973), is the opposite of industrial productivity. It means "autonomous and creative intercourse among persons, and the intercourse of persons with their environment; and this in contrast with the conditioned response of persons to the demands made upon them by others, and by a man-made environment." Illich considers conviviality to be "individual freedom realised in personal interdependence and, as such, an intrinsic ethical value." He also proposed a prescription: "We need procedures to ensure that controls over the tools of society are established and governed by political process rather than by decisions by experts."

For Sclove, a first step would be to require not just an environmental but a social and political 'impact statement' for new technological projects. Drawing up a social and political impact statement is not a job for the existing cadre of experts, raised on debating cost, feasibility, safety, environment and national security (and often framing the decisions according to instrumental rationality), but for the citizens of the affected community. Public debates and voluntary trials, with citizen participation, could be organised: their mandate is to assess the social and political implications of the technology, not its 'marketability' or its 'imagineering' – the spin that the company, its advertisers or market researchers would like to put on it to encourage acceptance and consumption. Actual social experiments along these lines have been conducted in Denmark, the Netherlands, the U.K., Germany and, most recently, the U.S.A. These are the most recent beacons for social experiments (institutionalised in Denmark, and gradually in the Netherlands) indicating that public participation in technology development yields satisfactory results.

FROM IDEAL CITIZEN TO IDEAL COMMUNITY

Sclove's argument concerning the ideal citizen and community is radical for its simplicity and self-evident 'goodness': let Kantian morality and autonomy (freedom and respect of others) as well as (Maslovian) self-actualisation be your philosophy of life; to engender mutual respect, encourage face-to-face interaction within the community in which you live (here he refers specifically to physical locality). Local virtual communities could be a tool, but shouldn't be ends in themselves for, among other reasons, electronic mediation could "produce psychological distancing from moral consequences" – the argument made in connection with the scientific bureaucracy in place for the construction of the Atomic Bomb and the systematic destruction of European Jewry. Empirical research on new technology always should address this notion of psychological distance from moral consequences.

To the greatest possible extent, the communities should be small, otherwise distance and numbers upset the building of mutual respect and understanding – the foundation of commonality, fruitful dialogue, and democratic decision-making. This is the ideal of autonomous, self-governing communities, which the Unabomber, too, expounded, along with the value of freedom. In revolutionary, neo-Luddite style, he wrote, however, that "revolution is easier than reform", and argued that "factories should be destroyed, technical books burned..." His brand of empowerment was destructive, his acts morally reprehensible.

Politically, Sclove's message is to develop local organisations and institutions which respect and further "globally aware egalitarian decentralisation and federation".

Centralisation implies concentration of power. In a strong democracy, power, ideally, is equal and diffused. Communities themselves should set agendas and work them through in the newly developed organisations and institutions. Everyone may contribute especially when basic policies are under discussion, though the citizens needn't contribute all the time. There is a middle ground between representative and hyper-direct democracy – between the functioning expertocracy and these bad dreams of everyday line-item teledemocratic voting we occasionally read about.

In terms of work, the human overseer in old factories and the (electronic) monitoring devices in present-day offices are the obvious figures of authority, direct means consciously used "to undermine conviviality and interorganisational solidarity". It's a vicious circle, for resistance often has provoked the tightening of authority. Automation (together with monitoring) breaks the ability to resist. Moreover, authoritarianism can be designed into the workplace, as in the Church (with the preacher on high at the lectern) and in the school room (with the teacher at the podium or at the large desk and the pupils at the smaller desks). Consider the private corner executive offices with windows, together in the same building with the open floorplan desks of secretaries and subordinates, each within sight of the superiors walking along the balconies above, on translucent floors as Hollywood has it. The authoritarian designs can also be conditioning. To Sclove, such layouts reproduce an "uncritical acceptance of the inevitability of technological forms that mirror the social patterns." Convivial, cooperative workspaces flatten these hierarchies. A number of scholars have proclaimed the 'end of public space'. Sclove argues that "democratic places are where citizens gather naturally and informally for fellowship, solace, recuperation, release, diversity, surprise, ritual, performance and politics." He cites other authors' descriptions: 'civic nuclei', 'talk shops', 'great good places'.

Taking on the attitude of the Amish and of Sclove, we should study the politics of mall and airport design, perhaps best summed up in the phrase "there are no demonstrations in the mall" (or at the airport for that matter). Gathering space has been privatised. We should study turn-key, secure communities under constant private surveillance and movements in the direction of the 'smart neighbourhood' or 'smart city', not just 'smart buildings' (workplaces) and 'smart homes'. Most probably, they are coming soon to a neighbourhood, town, city, country near you, your friends, your parents.

Sclove is reviving the anti-authoritarian, countercultural Sixties, out of which grew the 'appropriate technology' movement. What were these technologies of freedom? The bicycle, decentralised solar energy, community windmills, methane gas and gasohol, mass transit, recycling, wood-burning stoves, and compost heaps, each more 'appropriate' than the centralised or radical individualist or unsustainable alternatives – nuclear power, oil, private cars or electric garbage disposal. Pursell argues that they were given up for their supposed affinities with femininity, in the post-Vietnam remasculinisation of the U.S., under Ronald Reagan and Rambo.(03)

Owing to the demise of the appropriate technology movement it makes sense that Sclove has looked abroad to bring back home Jeffersonian democracy, pointing to current expressions of democratisation of technology design, assessment and decision-making in Europe. Citizen Juries, judging new technocultural policy, may have been an innovation of the Jefferson Institute in Minnesota but it's in Denmark — which hasn't had the systematising, scientific tradition synonymous with Modernism and the divide between the expert and the lay person — where the incipient 'strong democracy' model is supposedly taking shape.

The Danish Technology Board, like the Rathenau Institute in the Netherlands (both of which were inspired by the now defunct American Office of Technology Assessment), came into being after the massive protests against nuclear power. As in the Netherlands, where technology assessment (of the negative effects of technology after the fact) has been expanded to include constructive technology assessment of the design prior to its release, the Danish Board has gradually expanded its methods to include more and more public participation. It practices the following: single expert analysis (for overviews on a subject), cross-discipline expert group analysis (when a technocratic solution is publicly and politically acceptable), participatory methods including scenario workshops (or future labs), consensus conferences (in order for the process to gain credibility for the political decision-makers) and public debates.

There is a Danish word — *folkeoplysning* — 'people's enlightenment', which provides a certain depth of meaning to public participation for the government and the people. The outcomes of the debate bear out this 'reverse enlightenment project'; experts are enlightened to that fact that citizens are well-informed and able to articulate nuanced opinion. They are appreciated: our familiar battles over the effects of new technologies — time-consuming, resource-intensive fights between lobby groups, companies and governmental ministries (often ending in court) — these are less familiar to Danes. These procedures empower the citizen to shape the outcome of debates on technology (and not just the fate of a product on the marketplace). In fact, not just the effects of technology (conventionally, the environmental concerns) but the design of technologies themselves are gradually being contested in public arenas with access to government.

The Rathenau Institute in the Netherlands also practices the constructive assessment of technologies in the making. The next step would be to assess visions of the future — 'technological vision assessment' — comparisons and debates on imagined futures, with particular technologies as the focal points. Are they authoritarian; do they promote radical individualism; are they convivial? The Dutch, too, have a much-loved word with a certain depth of meaning, which is a translation of 'convivial'. Wouldn't it make some sense to ask, sincerely, whether the proposed technology is *gezellig*?

(01)
Sclove, R., *DEMOCRACY & TECHNOLOGY*, Guilford Press, New York, 1995. The pages numbers in the text refer to Sclove.

(02)
On Amish decision-making concerning new technology, see also Reingold, H., "Look Who's Talking", *WIRED*, January, 1999.

(03)
Pursell, C., "The Rise and Fall of the Appropriate Technology Movement", *TECHNOLOGY & CULTURE*, 34, 3, 1993.

10

TRUTH IN THE MUSEUM & THE REFLEXIVE EXHIBITION

Truth lives on a credit system; our thoughts and beliefs pass so long as nobody refuses them, just as bank-notes pass so long as nobody refuses them. **(01)**

There is an expectation that truth is represented in the museum, and more specificially in those museums thought of as 'revered institutions'. Why has this expectation of truth in the 'revered institution' developed? Historically, cabinets of curiosities and museums emerged from within a segment of society which was considered to have a duty to speak the truth. Indeed they also were considered to have a superior capacity to generate truth. I speak, initially, of the gentle English society of the seventeenth and eighteenth century – those enlightened men of the 'better ranks of society', falling in between the yeomanry and the noble courts, and populated by men of means and enlightened pursuits – taxonomists, astronomers, geologists – leisure collectors and scientists. Unlike the ungentle yeomanry and unlike the aristocracy, it was only the 'gentle' who publicly and privately strove to maintain the ideal of secular veracity. Enlightened truth-sayers was their self-image, communicated throughout gentle society.

In Steven Shapin's *A SOCIAL HISTORY OF TRUTH* (1996), an important distinction is made (by Daniel Defoe) in the eighteenth century between two different types of truth-telling, which is still of great relevance today, also for museum exhibitors. It's the distinction between truth-telling by the gentleman and truth-telling by the tradesman. In the seventeenth and eighteenth centuries the gentleman was expected to tell the 'whole truth' (in some sense, he could afford it), but the tradesman was not (he could not afford it). Importantly, this did not make the tradesman dishonest:

> *[T]he tradesman's word and promise had to be taken in the context of the 'circumstances of trade'... When a tradesman swore that he would accept no less than a guinea for a piece of cloth, he knew that in fact he would accept less. These 'trading lies' were therefore literally untrue but contextually benign.* **(02)**

For some time now museums have been caught between the rock of gentleman's truth-telling and the hard place of tradesman's honest untruths. Having originally been patronised or funded by the gentleman, gradually the museum has given itself over to the tradesman, that is, the sponsor. This sponsor can be science and the patrons of science. It can be the state. It can be the corporation.

When sponsored by science, the truths that are conveyed in the museum therefore should be considered within the 'circumstances of science', with scientists being both gentlemanly and tradesmanly. When sponsored by the state, the truths represented in the museum must be considered within the 'circumstances of the national economy and national honour' and, when the corporation is the sponsor, within the 'circumstances of corporate trade and corporate honour'. I have added the word honour here in the sense of respecting the sponsor's truth-telling, for the sponsor is still treated in a gentlemanly manner, however much the balance has of his truth has tipped towards the tradesman's.

Not everyone accepts the fact that the museum exhibition must be considered within the context of the 'circumstances of trade' or, indeed within the 'circumstances of sponsorship'. The museum continues to be held to the standard of the gentleman's truth, not least by museum exhibitors themselves, not least by the museum visitor and indeed not least by the museum exhibition critic.

Where the museum curators are concerned there is the standard museum policy (communicated, for example, by the Science Museum, London) that the sponsor cannot bias the content of the exhibition. Thus gentlemanly, not tradesmanly, truth will be upheld. The institution derives its gentle reverence therefore not only from its history but from present-day policy. Such a gentle appearance of truth-telling opens it up for critical examination.

Sponsored exhibitions remain problematic — to deal with the 'circumstantial' truths built into them a reflexive approach could be employed resulting in a 'reflexive exhibition'. It is not the answer to the problem, but one answer, with its own set of dilemmas. From a museological point of view this essay could fall within the realm of exhibition epistemology, or how to exhibit senses of truth. My perspective is sociological and historiographical, reflecting the increasing interest social scientists have taken in representations of science, technology and culture in the press, on television, in cinema and advertising — the various forms of popular technological culture.

The museum has long been a populariser of science but as the revered institution, it cannot be classified so easily as a 'science populariser', if popularisation connotes 'selling science' by the tradesman. The revered institution, conversely, maintains the appearance of representing the gentleman's truth. An indication of how the museum may introduce and frame an exhibition as if it presented a kind of tradesman's truth, to be seen within the 'circumstances of sponsorship' can be seen in the following examples. It also points to the dilemmas present in adhering to and being held to the standard of gentlemanly truth, apparent in the notion of the museum as 'revered institution'.

The first example derived from the early twentieth century gentleman's world of New York City, as it still exists according to a Californian feminist. In 1984 Donna Harraway, the philosopher and sociologist of science, published an article entitled "Teddy Bear Patriarchy"; an analysis of the African Hall at the American Museum of Natural History in New York City (a private institution).[03] The groundwork was laid by antebellum gentlemen, and has been added to by collector-scientists, museologists and their patrons.

'Teddy' refers to Theodore Roosevelt, the swashbuckling president, who, like many others in the white, protestant, capitalist middle and upper classes was "committed to nature, camping and the outdoor life". The article posits the gentlemen patrons as white, protestant and capitalist and helped usher in some of the main themes and identify some of the main targets of postmodern cultural studies — the revered gentlemanly 'truth-telling' institution being the one scrutiny.

African Hall promotes Teddy's huntsman outdoorsman life and much that is considered incorrect these days. Harraway's central theme is that the "main purpose of African Hall is not to afford an understanding of the natural world but to reproduce and naturalise a master narrative of race, class and gender in a capitalist society".**(04)** Harraway's is an origins narrative about ideological intent of how wealthy Male WASPS materialised versions of their truth in the exhibition halls. She explains how the purpose of the exhibits in African Hall is to promote a relevant version of vigorous evolution, against the backdrop of white racial decline, and the much debated talk of race suicide, whereby certain white races were outbreeding certain non-white races. One of the founders summed up the intention of the exhibits:

They all tend to demonstrate the slow upward ascent and the struggle of man from the lower to higher stages, physically, morally, intellectually and spiritually. Reverently and carefully examined, they put man upwards towards a higher and better future and away from the purely animal stage of life. **(05)**

Revered institutions may sponsor particular versions of gentlemanly truth.

Another museum setting, which analogously could be deconstructed as an ideological project, is the National Museum of Science & Industry in Vienna, opened in 1909, to exhibit the fruits of Austrian and German ingenuity. Until it was emptied recently, before a debate about whether it should be 'updated' or turned into a science centre, it had unmistakable overtones ripe for deconstruction (nationalism, racism, sexism and colonialism, for example). If museum curators sustain an ideologically inspired master narrative they are thereby 'implicated' in the history of suppression. This is the 'political act' of exhibition in the context of the revered institution.

To make a comparison of an exhibition at a company, or tradesman's, museum and an exhibition at a revered institution, or gentleman's museum, I refer to the Eurotunnel Exhibition Centre in Folkestone, England (now closed) and to the Science Museum, London.**(06)** These are exhibition spaces, examined in 1993 and 1994 respectively, devoted to a new technological project, the Channel Tunnel. The Science Museum exhibition was sponsored by Eurotunnel, but the sponsorship followed guidelines about not biasing the content, which I take to mean that the revered institution would not incorporate the tradesman's version of the truth.

While newspapers in Britain were churning out shocking stories about how the company's 'commercial objectives' were interfering with 'public safety needs', Eurotunnel's Centre didn't bother to depict or explain how the system had been designed with safety in mind.**(07)** In the tradesman's museum – a 'museum' for its display techniques not for its gentlemanly reverence – not a single safety issue or potential tunnel disaster was addressed.

Instead adult doll figures with television screens for heads tell us, in children's voices, that the experience will be fun for the whole family.

What of the gentleman's museum? The Science Museum in London opened an exhibition to coincide with the opening of the Channel Tunnel in 1994. The following quotation is from a newspaper article on the exhibition, which reveals the extent to which safety issues and disaster scenarios dominated the British public debate about the tunnel as well as the exhibition.

> *The first thing confronting visitors to the Science Museum's new exhibition...*
> *is a flickering television screen showing nine headlines: Power Failure, Flooding,*
> *Suffocation, Tunnel Collapse, Explosion, Rabies, Railway Accidents, Fire and Train*
> *Breakdown. Yes, these are all things which might happen... in the Channel Tunnel.*
> *It's enough to make you want to turn and run, even in Central London. It's certainly*
> *no inducement to go all the way down to the coast to try the real thing.*
>
> *But the exhibition organiser... insists there is a method in this apparent madness.*
> *"We thought it was important to confront people's fears. These are the most*
> *common reasons people gave in a survey for not wanting to use the tunnel.*
> *I think it will be like flying – some people will always be frightened. You can't*
> *do much about that."* **(08)**

Later in the article, a Eurotunnel spokesperson is quoted as saying, "People often express a fear of the unknown." In Enlightenment or Modernist terms, fear is equated here with superstition and once truth is understood, there is no need to fear, to be be superstitious.

Touching the disaster pictograms the hypertext program would display a screen explaining Eurotunnel's safety system design solutions. There is nothing to fear, in other words. I will not go into the adequacy of that kind of explanation for fear, nor the proposed design solutions to take it away.**(09)** Instead I would like to touch on what could be called the 'circumstances of sponsorship' in connection with the version of truth in the exhibition. Then I'll put forward some lessons.

Usually the sponsor is considered to be the expert on the issue, and Eurotunnel undoubtedly has expertise in building the tunnel and communicating the tunnel to the public. In the case of the Science Museum, the curators first went to the sponsor-expert, drew up a storyline and then had the facts verified by outside experts to avoid bias. Facts, however, are not innocent, especially when they are arranged in order to tell a story. In fact the exhibition presented Eurotunnel's story – a kind of Eurotunnel enlightenment project.

As standard practice Eurotunnel had a sophisticated risk assessment performed on the tunnel design, and presented the document to the regulators in order to convince them the system would be safe. The risks that Eurotunnel took into account in their analysis were the same and only risks depicted in the exhibition. Eurotunnel's design solutions were popularised in the exhibition; more specifically, the risk assessment was popularised in the museum.

According to opinion polls (to which the curators makes reference) it was the prospect of terrorist attack that troubled Britons most about the tunnel. But very little public information is available on how Eurotunnel and police authorities would deal with that risk. It is also not dealt with in the risk assessment document; that information — including those design and policy solutions — are confidential. In turn, the risk and fear of terrorist attack is not in the museum exhibition. Something else is also missing, but for a different reason: claustrophobia, a fear that television newsmakers and the ferry companies made great play of, is absent. Why? Because claustrophobia is in the ferries' storyline about fears of using the tunnel, not Eurotunnel's (risk assessment).

Another example of the 'circumstances of sponsorship', or the exhibition's 'non-sponsorship' by the ferry companies, relates to corporate competition. Eurotunnel advertised their system as faster — thirty five minutes on Le Shuttle via the tunnel; an hour on a conventional ferry and a little less on the hovercraft. To Eurotunnel, the future customer was after speed and speed was the selling point. The ferry companies, however, claim better on-board amenities and that the boat trip is part of the holiday, that when the weather is fine it's far more pleasant to take the ferry than to go through a hole in ground.

To depict the competition between the ferries and Eurotunnel in the museum the curators pose a question, answered by piling foam blocks onto each other in order to compare the different modes of transportation. The thickness of the foam blocks indicates the speed it takes to cross. The basis of the comparison, in other words, is speed. And what question do the curators pose? "The Best Way to Cross the Channel?" 'Best' here means fastest.

This essay is entitled "Truth in the Museum and the Reflexive Exhibition", and it's a normative point. What I mean is that the curator has a subtle, social-scientific task to perform in thinking through the arrangement and interpretation (the frame of reference). To determine whether the frame of reference is that of the sponsor the curator must take the time to 'map the public debate', determining the frames of reference of the other major parties involved. In the Science Museum exhibition, the ferry companies' were ignored completely, skewing the interpretation of the tunnel in a gentleman's museum in favour of Eurotunnel's tradesman's truth-telling.

In scholarship, reflexivity means turning the tools used to analyse an object of study onto yourself. This allows you to take into account the effects of the observer, it allows you to determine whether your own 'normative commitments' are framing your findings. Clearly, the curators in London agreed with Eurotunnel's assessment idea that people must be 'fearing the unknown'. Do the curators 'fear the unknown'? What do the competitors, independent experts and science and technology analysts 'know' about the tunnel, quite apart from Eurotunnel's tradesmanly knowledge? Do they 'fear the unknown'?

Normative commitments, inherited from the sponsor, may frame the narrative. The reflexive exhibitor would state this unequivocally within the narrative and not merely on the acknowledgements board. As promised, I wish to conclude with the central dilemma of the reflexivity thesis. By admitting the 'circumstances of trade' in museum truth constructions, the exhibitor could very well undermine the revered institution's credibility, in the eyes of those who still believe in gentleman's truth.

This essay is adapted from a public lecture given in the series, *BETWEEN SCIENCE AND THE PEOPLE*, newMetropolis Science & Technology Center, Amsterdam, September, 1997.

(01)
Shapin, S., *A SOCIAL HISTORY OF TRUTH*, Univ. of Chicago Press, Chicago, 1995.
(02)
ibid. 1995, p. 95.
(03)
Harraway, D., "Teddy Bear Patriarchy", in *PRIMATE VISIONS*, Routledge, New York, 1989.

(04)
This summary comes form a critical review of Harraway's article. Schudson, M., "Paper Tiger", *LINGUA FRANCA*, August 1997, pp. 49-56.
(05)
Harraway, D., "Teddy Bear Patriarchy", 1989, p. 57.
(06)
See Rogers, R., "Do You Want to Go for a Ride on the Chunnel?", *PUBLIC UNDERSTANDING OF SCIENCE*, 4, September 1995, pp. 363-396; and Rogers, R., "Managing British Public Opinion of the Channel Tunnel", *TECHNOLOGY & CULTURE*, 36, 3, July 1995, 636-640.

(07)
DAILY EXPRESS, 14 May 1994.
(08)
See Rogers, R., *ENGLAND & THE CHANNEL TUNNEL*, Ph.D. dissertation, University of Amsterdam, 1998, chapter 6.
(09)
See, for example, Fuller, S., Philosophy, *RHETORIC & THE END OF KNOWLEDGE*, Univ. of Wisconsin Press, Madison, 1993.

DESIGN FOR ALL: TECHNOLOGIES FOR THE SITUATIONALLY CHALLENGED

THE essay concerns evolutions in thinking about cultural and technological design, and for whom that design is intended. It concerns the roads taken, especially in the United States, towards acknowledging and building for 'the other': the biblically meek or needy, the dependent, the special, or the variously challenged, as people with disabilities have been viewed and called, judgementally, over the past one hundred years.

Conventionally, Design for All, the new design moniker, could be thought of as a rallying cry for a social movement. A movement calling for attitudinal and structural change in society. A movement with a history. Obvious sources of guidance and inspiration for such a movement across Europe may include successful barrier-free movements in the United States. In the U.S. the movement's rallying cries and legislative successes have benefited from the civil rights movement, from calls to end notions of 'separate but equal', to end conscious or unconscious segregation, still the case at many private and public institutions. The movement has its symbolic beginnings, its seminal moments in history, such as in Denver in 1978, the first major public demonstration by wheelchair users against inaccessibility. (To press the analogy with the civil rights moments, we could call it a Rosa Parks moment in history for the disabled or the differently abled.)

Over the years the scope of anti-discrimination laws has broadened. In the U.S. the movement culminated initially in federal anti-discrimination laws for the handicapped in 1973, some five years before the Denver demonstration. It culminated again in 1983 with the creation of a dedicated federal agency – the National Council for Persons with Disabilities – and more recently with the Americans with Disabilities Act, signed into law in 1990: "The Americans with Disabilities Act prohibits discrimination on the basis of disability in employment, programs and services provided by state and local governments, goods and services provided by private companies, and in commercial facilities."

After the 1990 act architectural and communication barriers would have to be removed across the land. In fact a January 1992 deadline was set for 'readily removable' barriers. Breaking down barriers, the movement's rallying cry, while recognised by federal law, is still being made at the grassroots level, with many of the well-known trappings: symbols, slogans, poetry and song.

So much for the recent history of the movement. Before we become somewhat philosophical, I'd like to provide you with my own perspective on the history of the movement and the different consciousnesses which have informed it. Though such histories must be somewhat artificially constructed, for the sake of clarity I would like to trace three overlapping generations of American consciousness towards the disabled, at least as I experienced them through my father, who worked in the aftermaths of the first and second generations, and who modestly inspired the third as the executive director of a New Hampshire rehabilitation and outplacement centre for the disabled – the Debbie Voter Center in Salem, New Hampshire. I learned of these generations on the back porch in New England, my place for deeper reflection.

The three overlapping generations of American consciousness towards the disabled may be summarised in these words: pity, dependence, and independence. The first is the charity model, embodied in one of the world's most famous advocacy organisations for both the mentally and physically disabled: Goodwill Industries. It was founded in New England, in turn-of-the-century Boston, by a Methodist minister. It was and to some extent still is a kind of Salvation Army model, though without the redemptive, militaristic and missionary overtones. As they still do today, Goodwill would solicit and receive donations for the good cause. These donations are the things we discard, old clothes and household items. In the Goodwill factory workshops, the disabled would be employed to sift through and sort the old things, clean them up, for sale at the Goodwill stores and auctions, in order to collect money to fund further work and training largely of this kind.

Putting it rather unflatteringly, one could say that the materially-discarded and the otherwise socially-discarded were brought together under one roof, a kind of factory for and of the disabled. It was a factory-charity. Your good will would be expressed in your donations to fund the infrastructure behind work schemes. The disabled became or were made useful. Work would provide dignity, give people something to do.

In this sense it was an 'industrial program', as the founding Minister called it. I still vividly recall the long tables chockablock full of donations and the occasional trash bag gone astray, all being sifted by disabled workers, at the Goodwill in Lowell, Massachusetts. There also were small assembly lines for mass-production-style piece-work. Labour was cheap, and so were the products. Wires would be strung through the stems of green carnations for Irish festivals, springs would be loaded into spinning tops for under the Christmas trees, in a kind of Taiwanese factory on the Merrimack River. All manner of doo-dads and thingamajigs were assembled and shipped to the wholesalers for sale to the retailers. This was low IQ work for the lesser and greater endowed alike. Sometimes the able-minded disabled would supervise on the assembly line floor; the able-bodied and able-minded would manage in the offices.

The dependency attitude – our second overlapping generation – is embodied in the disability insurance model, or even the Veteran's Administration model (however much civilian and veterans' programs always have been separated administratively). Simply put, disability benefits compensate for physical loss, or physical loss and wage loss, or physical loss, wage loss and wage-earning capacity loss. As at least one analyst has put it, one was treated as a pensioner, paid to retire and remain retired, no matter your age.**(01)**

In practice, what the first two generations have in common (according to Berkowitz) is that "people [have] perceived the handicapped as less intelligent, less able to make the right decisions, and therefore less able to determine their own lives than the non-handicapped". At Goodwill, the mentally and physically handicapped were together under one roof, not so much unlike the situation at the dreaded custodial institution, state hospitals, often thought of as a 'disabling environment' for the disabled.**(02)** These places reinforce disability, however well-intentioned and well-aware they have been of criticism. The overly paternalising treatment of such an institution is not forgotten today. One typical slogan is "we're not dead yet".

The closure of U.S. state institutions, in the 1950s and 1960s, marked a turning point from custodial to 'community care'. While the models of the first and second generations would remain in place, because they've often moved with the times somewhat and have been financially successful (Goodwill is a $1.2 billion organisation), a third model – independence – would gradually emerge.

In my youth in New England I learned about the group home concept, the forerunner to independent living. My father set up one of these communities, smaller but not so much unlike Het Dorp, in the Netherlands, and more recently the Paralympic Village in Atlanta. Planned utopias for the disabled, where all is accessible, as long as one remains within the confines of that area. Mobility was guaranteed, though unlike the ultimate goal of Design for All, the disabled or the wheelchair user was hard-pressed to happily leave that well-known blue sign. One's address remained identifiable and stigmatising, one's mental and physical confines limited.

While Britain seems to be returning to the custodial model (*INTERNATIONAL HERALD TRIBUNE*, 14-15 February, 1998), in the U.S. the call is for independent living, also through assistive technology. Full social involvement and regular jobs outside of the sheltered workshops of the Goodwill model. Regular living, outside of the group homes and the mini-utopias designed for the disabled. Beyond the blue sign, even. Real progress will have been made when all those wheelchair signs have become redundant.

- The notion of mainstreaming is strangely embodied in the Barbie doll in a wheelchair, Share a Smile Becky. Once it was realised that Becky did not fit into the elevator of Barbie's Dreamhouse, the Dreamhouse (not the doll or the wheelchair) was redesigned.

FROM SOCIAL MOVEMENT TO TECHNOCULTURAL MOVEMENT

Design for All is better conceived not as a social movement but as a technocultural movement, a new voice calling for change in both the culture of technological design and our technological culture.

The idea of 'our technological culture' first recognises and then exploits the absence of nature as we conventionally conceive it. Metaphorically the Netherlands has become a cross between an airport, a greenhouse and a park – all very much man-made. It's a manufactured nature; some are calling it 'new nature'. The Netherlands has become a designed landscape. Nature has become technoculture.

Once we take on board the notion of the 'Netherlands as designed landscape', a designed technocultural landscape, devoid of nature, opportunities for further design are implied, opportunities to be exploited. If it is a design-scape, then it could, in principle, be designed for all. I'm not sure certain environmentalists would appreciate more mobility and face-to-face interaction for all. There are some who advocate travelling without moving, the

Wellsian idea of the time and space traveller who remains in his high-tech armchair, roaming far and wide without ever really leaving his home and his chair. At least that's the interpretation of the Wellsian piece, passed on to Frank Herbert's *DUNE* and then to Jamiroquai's music, in the age of the Internet, in the age of telework and teletravel. Not too many of the 'travelling without moving' school realise that if motion is restricted to the virtual realm they'd be knocking down the ladder for the 'mobility-challenged'.

For the time being, our landscape remains designed for motion. Generally speaking, this landscape is continually being designed and redesigned for optimal movement, flow, transportation, throughput and distribution, within certain space and environmental constraints. The airport model, I suppose we could call it. Keeping in mind the Netherlands as design-scape, we could say that yesterday's 'landscape architects' are today's traffic engineers or, if you will, 'mobility manufacturers'. We can't get enough of it. This is not that surprising; indeed Aristotle called purposive mobility one of the central characteristics of higher life forms. Communication also was on his list. We must move and interact, meaningfully. In the stream of life. The mainstream of life.

Perhaps it's time to discuss and take on board the notion of a country as design-scape, built for motion, in order to create not segregated mini-utopias, but a heterotopia (the new term), which strives to accommodate (not force!) different abilities and mobilities by design.

I'd like to mention the body, a mainstream body, another 'nature' or natural product, at least as we conventionally think of it. If, by introducing 'new nature', we rethink and deny the existence of nature as it is conventially conceived, then we may also, rather plausibly, rethink and speak of the absence of the natural body. Think of all the rehabilitative and cosmetic surgery, all the prostheses. That is, in principle, the body may be redesigned. Indeed, the body may be redesigned to fit the landscape.

1978 may have been the symbolic beginning of Design for All-type movements, in Denver. The Seventies, however, also witnessed the bionic man and woman. Prior to and throughout the Seventies the emphasis was on designing out disability, or designing bodily ability. In the United States in the 1920s, for example, it was the official policy in certain companies that one's disability be hidden. In automobile factories it was considered off-putting and poor for worker morale if the injured or disabled stuck out. It also gave a big boost to manufacturers of body part replacements.

Once disability is defined as both a social and medical problem, body redesign often is implied. Where recent technologies as the cochlear implant are concerned – some have called it the bionic ear – we clearly have a case of redesigning the body to fit the existing landscape. Those who wish to redesign the body to fit the landscape, could be called the adherents of an 'anti-design for all', a counter-movement, a school of 'redesign of all'. Herein lies the central dilemma: Redesign the body to fit the landscape, or redesign the landscape to fit the body?

The subject to ponder therefore is the choice between 'Landscape Redesign for All', and 'Body Redesign of All', especially if 'Design for All' becomes recognised certification, like the 'green point' for environmentally friendly products in the Netherlands and Germany. Should a bionic ear be considered a 'design for all' product? No, it's a redesign of all.

'Design for All' is not only a rallying cry for a change in our technological culture but a compelling aphorism for designers — short, snappy and memorable — a guiding vision for every designer, every institution dedicated to design. The sort of call which could glue an organisation together, in the same way the notion of 'universal service' once rallied the engineers and management at AT&T.

The culture of technological design is embodied in institutions such as the Netherlands Design Institute and the Royal College of Art; they are front stages, revered theatrical spaces with performers and new props or products to dazzle — to capture the imagination, to create momentum, to mobilise people and institutions, to secure funding. To do so, the performers or designers must believe in their own ability to effect positive technocultural change. They need big ideas summarised in short-hand expressions. Design for All is compelling because it implies that everyone is challenged, everyone is 'crooked'.

Here are some examples of heterogeneous or heterotopic product design for the crooked mainstream: in the U.S. there is a television service for the hard of hearing, 'captions for the hearing impaired'. In Washington DC I noticed recently that café televisions come with these captions. Loud cafés challenge everyone's ability to hear; providing subtitles on café televisions is design for all.

Another example occurred to me when I visited the Smithsonian Museum of Natural History and asked about the self-guided, interactive audio tours. The man looked me in the eye and said: "Take the Blind Tour!" Not just for the blind, this ostensibly 'special needs' technology was mainstreaming.

The final example is a beach wheelchair I came upon in Rehobeth Beach, Delaware. It looks like a lot of fun for anyone and I saw a series of the 'non-vertically challenged' using it. Nowhere did I see a blue wheelchair sign — the chair had left the sign.

Instead of continually redesigning bodies or retrofitting dreamhouses, we should take up the challenge of making products and technocultures which accommodate the 'situationally challenged' (as in that café or on that beach). Design problem definitions and proposed solutions could come from home, work and street 'situations', from everyday experiences, from the poetry of the heterogeneous mainstream: the crooked timber of humanity (Immanuel Kant & Isaiah Berlin). This would be design by all, a third generation technology, recognising that 'special needs design' per se is a thing of the past and that 'design for all' products, like their users over the last two decades, are now mainstreaming.

This essay is adapted from the annual *SUZE RIETDIJK LECTURE*, organised by the Platform Design for All, and given at the Netherlands Design Institute, Amsterdam, December, 1997.

(01)
Berkowitz, E., *DISABLED POLICY*, Cambridge University Press, Cambridge, 1987.

(02)
Swain, J., V. Finkelstein et al., *DISABLING BARRIERS – ENABLING ENVIRONMENTS*, The Open University, Sage, London, 1993.

SEE ALSO
Albrecht, G., *THE DISABILITY BUSINESS*, Sage, Newbury Park, 1992.

Berkowitz, E., "Domestic Politics and International Expertise in the History of American Disability Policy", *THE MILBANK QUARTERLY*, 67, 2, 1, 1989.

Berlin, I., *THE CROOKED TIMBER OF HUMANITY*, Vintage, New York, 1992.

Blaxter, M., *THE MEANING OF DISABILITY*, Heineman, London, 1976.

Blume, S., "The Rhetoric and Counter-Rhetoric of a 'Bionic' Technology", *SCIENCE, TECHNOLOGY & HUMAN VALUES*, 22, 1, Winter, 1997.

Borgman, A., "Information and Mobility", paper presented at the Mobility Conference, University of Maastricht, May, 1997.

European Commission, Userfit, Tide, ECSC-EC-EAEC, Brussels-Luxembourg, 1996.

Office of Technology Assessment, Technology and Handicapped People, summary, Washington, DC, May, 1982.

Oliver, M., *THE POLITICS OF DISABLEMENT*, Macmillan, Houndmills, 1990.

Rogers, R., "Take the Blind Tour! And 'Other Technologies for All'", *EASST REVIEW*, 16, 3, September, 1997.

Sarlemijn, A. and H. Boddendijk (eds.), *QFD*, Boom, Amsterdam, 1995.

Tanenbaum, S., *ENGINEERING DISABILITY*, Temple University Press, Philadelphia, 1986.

Topliss, E., *SOCIAL RESPONSES TO HANDICAP*, Longman, London, 1982.

Wagner, G., "Technopoetry", *EASST REVIEW*, 16, 3, September, 1997.

Royal College of Art
Postgraduate Art & Design

RCA CRD RESEARCH

CRD DOCUMENTS: A SERIES OF POLEMICS, MANIFESTOS, ESSAYS AND LECTURES ON COMPUTER RELATED DESIGN

TECHNOLOGICAL LANDSCAPES recasts the history of technocultural ideas through a series of essays exploring the visions, ideals and ideologies informing the design and choice of new technology-from the railways to the Internet, from the speed culture to the technological sublime. The essays gradually develop a critique of technology promotion, inviting designers and developers to rethink the language of technology.

By demonstrating the force of historical analogy for devising new cultural agendas, *TECHNOLOGICAL LANDSCAPES* participates in the politics of expectation, and the redesign of contemporary technological culture.

Dr. Richard Rogers lectures at the Royal College of Art, London and the University of Amsterdam.

ISBN 1-874175-28-4   £8.99

TECHNOLOGICAL LANDSCAPES
RICHARD ROGERS